Davenport–Zannier Polynomials and Dessins d'Enfants

Mathematical
Surveys
and
Monographs

Volume 249

Davenport–Zannier Polynomials and Dessins d'Enfants

Nikolai M. Adrianov
Fedor Pakovich
Alexander K. Zvonkin

American
Mathematical
Society
Providence, Rhode Island

2010 *Mathematics Subject Classification.* Primary 11G32;
Secondary 05C05, 05C10, 05C22, 05E18, 11R32, 12E10, 20B15.

For additional information and updates on this book, visit
www.ams.org/bookpages/surv-249

Library of Congress Cataloging-in-Publication Data

Names: Adrianov, Nikolai M., 1973– author. | Pakovich, Fedor, 1970– author. | Zvonkin, A. K. (Aleksandr Kalmanovich), 1948– author.

Title: Davenport–Zannier polynomials and Dessins d'enfants / Nikolai M. Adrianov, Fedor Pakovich, Alexander K. Zvonkine.

Description: Providence, Rhode Island : American Mathematical Society, [2020] | Series: Mathematical surveys and monographs, 0076-5376 ; volume 249 | Includes bibliographical references and index.

Identifiers: LCCN 2020012961 | ISBN 9781470456344 (softcover) | ISBN 9781470460297 (ebook)

Subjects: LCSH: Dessins d'enfants (Mathematics) | Arithmetical algebraic geometry. | Trees (Graph theory) | Galois theory. | Polynomials. | Algebraic fields. | AMS: Number theory – Arithmetic algebraic geometry (Diophantine geometry) – Dessins d'enfants, Belyĭ theory. | Combinatorics – Graph theory – Trees. | Combinatorics – Graph theory – Planar graphs; geometric and topological aspect | Combinatorics – Graph theory – Signed and weighted graphs. | Combinatorics – Algebraic combinatorics – Group actions on combinatorial structures. | Number theory – Algebraic number theory: global fields – Galois theory. | Field theory and polynomials – General field theory – Special polynomials. | Field theory and polynomials – Computational aspects of field theory and polynomials – Computational aspects of field theory and polynomials. | Group theory and generalizations – Permutation groups – Primitive groups.

Classification: LCC QA613.2 .A37 2020 | DDC 511/.5–dc23

LC record available at https://lccn.loc.gov/2020012961

∞ The paper used in this book is acid-free and falls within the guidelines
established to ensure permanence and durability.
Visit the AMS home page at `https://www.ams.org/`

10 9 8 7 6 5 4 3 2 1 25 24 23 22 21 20

Contents

Preface

Let $A, B \in \mathbb{C}[x]$ be two coprime polynomials, $\deg A = 10$, $\deg B = 15$, and consider the difference $A^3 - B^2$. Both A^3 and B^2 are of degree 30. How many higher coefficients of $A^3 - B^2$ can be made equal to zero? It is trivial to make the leading coefficient of $A^3 - B^2$ vanish; it is also simple to make the second coefficient disappear. Can we do more than that? What is the minimum degree of the difference $R = A^3 - B^2$ that can be attained?

It turns out that $\min \deg R = 6$, which means that, by a clever choice of A and B, we can make 24 higher coefficients of the difference $R = A^3 - B^2$ vanish, and it is impossible to do better.

The question, in a general form, concerning the pairs of polynomials (A, B) such that $\deg A = 2k$ and $\deg B = 3k$, so that $\deg A^3 = \deg B^2 = 6k$, was raised in 1965 [**BCHS-65**]. The above example corresponds to $k = 5$. For this case there exist four essentially different solutions (A, B) (the exact definition of what does it mean to be "essentially different" will be given later). For two of them, the coefficients of the polynomials are rational; for the other two, they belong to an imaginary quadratic field. It is incredibly difficult to compute these solutions. All the four pairs (A, B) were found only 40 years later. One of the solutions, defined over $\mathbb{Q}$, was already computed in 1965 in the original paper [**BCHS-65**]; the second one, also over $\mathbb{Q}$, was found 35 years later in [**Elk-00**]; finally, the pair of solutions over a quadratic field was found in 2005 in [**Shi-05**]; it turned out the the field in question (to which the coefficients of the polynomials belong) is $\mathbb{Q}(\sqrt{-3})$.

But there exists a wizardly method of proving all the above statements (except the fact that there is -3 under the square root) without any computation. This method consists in drawing very simple pictures: so simple that they are usually called *dessins d'enfants*. The latter means, in French, "children's drawings". This half-joking term was coined by Alexandre Grothendieck in his unpublished notes "Esquisse d'un Programme" (1984). Later on, these notes were published in [**Gro-84**] (both the French original and an English translation). Based on an earlier work of Belyĭ [**Bel-79**], Grothendieck pointed out that there are profound relations between (a) Galois theory; (b) Belyĭ functions (meromorphic functions with at most three critical values); and (c) a class of pictures drawn on Riemann surfaces. Nowadays, the theory of dessins d'enfants is an active (and, we dare say, fashionable) domain of research. The books [**JoWo-16**] and [**GiGo-12**] are dedicated to this subject. A vast bibliography is collected in [**SiVo-14**]. The book [**BoSt-04**], as well as our book, may be considered as particular chapters of this theory. The French expression *dessins d'enfants* is also currently used in English language literature. It has also been made into one of the entries of the AMS Subject Classification Index: 11G32.

Our book is among the most elementary ones dedicated to the subject: all our dessins are planar and "tree-like", and the meromorphic functions in question are just ordinary rational functions. However, this elementary character notwithstanding, the book preserves the most attractive aspect proper to the entire theory, namely, an intricate entanglement of combinatorics, the theory of polynomials, symbolic computations, special functions, Galois theory, number theory, and group theory.

The general class of polynomials we study in the book is as follows: they have the multiplicities of their roots fixed in advance. For example, the multiplicities of the roots of $P = A^3$ are all equal to 3 (or they may be multiples of 3), while the multiplicities of the roots of $Q = B^2$ are 2 (or multiples of 2). In the general setting, the root multiplicities will form a pair of partitions $\alpha, \beta \vdash n$ of the number n which is the common degree of the polynomials P and Q:

$$\alpha = (\alpha_1, \alpha_2, \ldots, \alpha_p), \quad \beta = (\beta_1, \beta_2, \ldots, \beta_q), \quad \sum_{i=1}^{p} \alpha_p = \sum_{j=1}^{q} \beta_j = n.$$

When, for a given pair of partitions (α, β), the degree $\deg(P - Q)$ attains its minimum, we call the pair of polynomials (P, Q) a *Davenport–Zannier pair* or, in short, a *DZ-pair*. There are many researchers who contributed to the study of these polynomials, so we might as well call them Birch–Chowla–Hall–Schinzel–Davenport–Stothers–Boccara–Zannier–Beukers–Stewart... polynomials (maybe the readers will be kind enough to add our own names to the list?), but for the sake of brevity we have chosen the two names which seem to us the most appropriate. The pair of partitions (α, β) is called the *passport* of the corresponding DZ-pair.

We call the dessins used in this book *weighted trees*. Such a tree should in fact be called *weighted bicolored plane tree*. It is an object like the one shown in Figure 0.1.

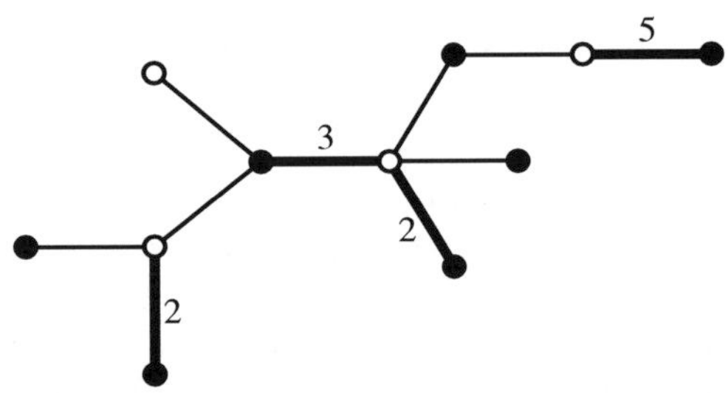

FIGURE 0.1. A weighted bicolored plane tree. The weights which are not explicitly indicated are equal to 1. The passport of this tree is $(5^2 2^3 1^2, 7^1 6^1 4^1 1^1)$.

Namely, it is a tree whose edges are endowed with *weights*, these weights being positive integers. The *degree* of a vertex is the sum of the weights of the edges incident to this vertex. We will also use the term *valency* as a synonym of the term degree. *Bicolored* means that the vertices are colored in black and white in such a way that the ends of each edge have opposite colors. *Plane* means that the cyclic order of branches around a vertex is taken into account: by changing this order we usually change the tree (though of course the new tree may turn out to be isomorphic to the initial one).

This notion will be explained in more detail later. The *passport* of a tree is the pair of partitions (α, β) representing the degrees of black, respectively white, vertices.

Let us, however, disclose our true intention: in fact, an edge of weight k represents a strand of k parallel edges: see Figure 1.1 (page 5). Thus, weighted trees represent a specific class of plane maps. It just happens that technically it is much easier to work with trees than with maps, especially when the number of parallel edges is a parameter which may take arbitrary values.

For the sake of brevity we will call these objects *weighted trees*, or even just *trees*, omitting all the adjectives.

Weighted trees are interesting combinatorial objects in their own right and may give rise to various studies involving enumeration, bijections and so on. We are not completely foreign to this experience: see Chapter 11 where we present some enumerative results. But mainly, for us, weighted trees serve as a remarkably efficient tool for studying DZ-pairs of polynomials. There is a bijection between the (equivalence classes of) DZ-pairs and the (isomorphism classes of) weighted trees having the same passport. Thus, for example, in order to establish the existence of a specific DZ-pair it suffices to draw the corresponding tree. The trees are also helpful in the study of the Galois action on DZ-pairs, but the construction in question needs a more lengthy explanation. Just one example: if, for a given passport, the tree with this passport is unique, then the coefficients of both P and Q are (more exactly, can be made) rational.

The contents of the book are as follows.

Chapter 1 is introductory: we present formal and detailed definitions of the main notions and explain the relations between them.

Chapter 2 is a very brief introduction to the theory of dessins d'enfants.

In Chapter 3 we show that for every "valuable" passport there exists a tree having this passport. In general (that is, for arbitrary maps) the results of this kind may turn out to be very difficult and the conditions of "valuability" may be very intricate. The Euler formula is, of course, a necessary condition, but it is far from being sufficient. However, for the particular case of weighted trees, the condition, which is both necessary and sufficient, is very simple. Namely, the number of vertices must not exceed the upper bound possible for the trees. The proof of the existence theorem is also easy. At the end of this chapter we explain what can be done for the non-valuable passports.

A very short Chapter 4 contains, as is implied by its title, the recapitulation of what has already been done in the previous chapters, and sets the goals for the subsequent ones.

Chapter 5 is a difficult reading. We establish a complete classification of what we call *unitrees*, that is, trees which are uniquely determined by their passport. Like busy foresters, we plant some trees, graft upon them branches of other trees, choose roots, put weights at various places.... The proof takes almost 30 pages, but we did not find a simpler one. As we have mentioned above, all the unitrees are "defined" over the field $\mathbb{Q}$ of rational numbers. This means that the equivalence class of the corresponding DZ-pairs contains a pair with the coefficients in $\mathbb{Q}$. There exist ten infinite series of unitrees described by some integer-valued parameters, and ten sporadic unitrees which do not belong to any series.

The last section of this chapter contains an example (just one example!) of a series of quadratic orbits, that is, a parametric passport which gives rise to two different trees. The situation is much more intricate than one might suppose.

In Chapter 6 we compute DZ-pairs corresponding to the unitrees. We have already mentioned above how difficult it is to compute a DZ-pair for a given tree. It is incomparably more difficult to compute them for an infinite series. In this chapter we will encounter such topics as Jacobi polynomials, Euler Beta function, Padé approximants, hypergeometric series, etc.

The passport is a Galois invariant, but there are many other invariants as well. The most advanced of them all is the monodromy group of the ramified covering of the Riemann complex sphere by itself corresponding to the rational function $f = P/R$ where $R = P - Q$ and (P, Q) is a DZ-pair. In Chapters 7 and 8 we provide a complete classification of the *primitive* monodromy groups of such coverings. For the definitions and statements, see the text. Not all the DZ-pairs corresponding to these groups are yet computed. To the best of our knowledge, a complete classification of the primitive monodromy groups of coverings of the sphere by itself is not yet achieved. The infinite series are not described, and a complete list of sporadic cases is not presented (maybe it is too large to be written explicitly). There are, however, some partial results, like a complete list of the affine groups appearing in this context which is given in [**MSW-11**]. Our result may be considered a modest contribution to the subject.

Chapter 9 studies some other Galois invariants, like symmetry, composition, self-duality, etc. The most unusual of them, and difficult to detect, is what we call a "megamap invariant". A *megamap* is a dessin which represents the Hurwitz space for a family of coverings of the sphere with *four* ramification points. The property of a dessin to serve as a megamap for such a family is a Galois invariant. Unfortunately, we do not have an algorithm to verify if a given dessin is a megamap for a Hurwitz family; it works only in the opposite direction, from a family to the corresponding megamap. Our examples are found by a kind of "blind search".

Chapter 10 is a case study: a very beautiful one, in fact. Here we consider a particular set of two trees which almost always constitutes a Galois orbit defined over a real quadratic field. However, from time to time this set splits into two Galois orbits, both defined over $\mathbb{Q}$. It turns out that the splitting cases correspond to the solutions of a Pell equation, and are "explained" by these solutions since the latter ensure that the discriminant of the quadratic equation in question is a perfect square. Pell's equation was studied for more than two thousand years. It is amazing to see that it still can tell us something new in the XXIst century. This example also illustrates the fact that there is absolutely no hope to find an exhaustive set of combinatorial and group-theoretic invariants which would provide us with an "if and only if" criterion of membership of two dessins to the same Galois orbit. Up to now, Diophantine invariants of dessins d'enfants were not yet thoroughly studied. They certainly deserve closer attention.

Chapter 11 is devoted to enumeration of weighted trees. We did not push this subject too far since it does not belong to the mainstream of our interests.

The last (and very short) Chapter 12 formulates some problems for subsequent study.

A few words should be said about the motivation. The first impulse for the study of DZ-pairs had come from number theory, in particular from the question of how close to each other two integer powers can be. But gradually this theory acquired its own intrinsic interest. The worthiness of a mathematical theory is explained by its internal beauty, by an abundance of non-trivial examples, by the difficulty of its main results, an ingenuity of their proofs, and, last but not least, by an interplay of various branches of mathematics. All these aspects are vastly presented in the study of the Davenport–Zannier polynomials.

Acknowledgments

We are particularly indebted to Gareth Jones who proved Theorem 7.8 at our request. Several Belyĭ functions were computed for us by Yuri Matiyasevich, John Voight, and Alexander Vatuzov. We are also profoundly grateful to our respective institutions as well as to numerous seminars and conferences where we presented our results and always found a stimulating environment, interest and support. David Roberts and Peter Müller kindly sent us the slides of their talks [**Rob-15**] and [**Mul-15**]. Conversations with Yuri Bilu and David Roberts were very enriching.

Many years ago, Umberto Zannier sent to A. Z. his paper [**Zan-95**], and this was the starting point of our interest in the problems described in this book. The second impetus came from a talk given by Cameron Stewart at the University of Bordeaux in 2010.

We are also grateful to the unknown referee for some pertinent remarks.

CHAPTER 1

Introduction

In 1844, Eugène Charles Catalan conjectured that 8 and 9 are the only powers of natural numbers which are at a distance one from one another. The conjecture was proved by Mihăilescu [**Mih-04**] 160 years later, in 2004. (A complete exposition of this result can be found in the book [**BBM-14**].) But the question of a possible distance between integer powers and, more generally, between powers of polynomials, still remains of a current interest. By the way, the two problems are obviously related: if the polynomials in question happen to be with integral coefficients then we may substitute an integer value of the variable and get a pair of close powers of integers. Gradually, however, it became clear that the polynomial version of the problem has its proper interest mainly due to a rich variety of relations with other branches of mathematics: we have already mentioned some of them in Preface. So, let us start our journey.

1.1. What are Davenport–Zannier polynomials

Our story begins in 1965, when Birch, Chowla, Hall and Schinzel [**BCHS-65**] asked a question which later became quite popular:

PROBLEM 1.1 (Cube minus square). Let A and B be two coprime polynomials with complex coefficients. What is the possible minimum degree of the difference $R = A^3 - B^2$?

Indeed, a reasonable measure of a "distance" between two polynomials is the degree of their difference. Let us start directly with an example.

EXAMPLE 1.2 (Elkies, 2000 [**Elk-00**]). Let us take the following two polynomials of degree 30: P is a cube of a polynomial A of degree 10, and Q is a square of a polynomial B of degree 15.

$$
\begin{aligned}
(1.1)\qquad P &= (x^{10} - 2x^9 + 33x^8 - 12x^7 + 378x^6 + 336x^5 + 2862x^4 \\
&\quad + 2652x^3 + 14\,397x^2 + 9922x + 18\,553)^3, \\
(1.2)\qquad Q &= (x^{15} - 3x^{14} + 51x^{13} - 67x^{12} + 969x^{11} + 33x^{10} + 10\,963x^9 \\
&\quad + 9729x^8 + 96\,507x^7 + 108\,631x^6 + 580\,785x^5 + 700\,503x^4 \\
&\quad + 2\,102\,099x^3 + 1\,877\,667x^2 + 3\,904\,161x + 1\,164\,691)^2.
\end{aligned}
$$

At first glance, these polynomials have nothing in common. But let us look at their difference $R = P - Q$. It turns out that all the coefficients of R in front of the degrees from 30 down to 7 vanish, and what remains is a polynomial of degree 6:

$$(1.3)\qquad R = 2^6\,3^{15}(5x^6 - 6x^5 + 111x^4 + 64x^3 + 795x^2 + 1254x + 5477).$$

Incredible, isn't it? We will return once again to this example in Section 9.5, page 146.

If you have Maple or another system of symbolic computations at your disposal, then the verification of the above equality is trivial. It is even feasible, though tedious, to carry out such a verification by hand. Is it also simple to compute the polynomials themselves?

EXERCISE 1.3 (Trying to compute the above polynomials). Let us try a frontal attack. Take your favorite system of symbolic computations; write down polynomials $A(x) = \sum_{i=0}^{10} a_i x^i$ and $B(x) = \sum_{j=0}^{15} b_j x^j$ with indeterminate coefficients. Let $R = A^3 - B^2$. Compute explicitly the coefficients of $R(x) = \sum_{k=0}^{30} c_k x^k$ and equate the coefficients $c_7, c_8, \ldots, c_{30}$ to zero. You will get a system of 24 algebraic equations with $11 + 16 = 27$ unknowns. Following Elkies's result, set the leading coefficients $a_{10} = 1, b_{15} = 1$, and set also $a_9 = -2$. What is the degree of the system thus obtained?

SOLUTION. The system takes a couple of pages. It contains 18 equations of degree 3, four equations of degree 2, one equation of degree 1, and one equality which is satisfied automatically ($c_{30} = 0$). Therefore, the total degree of the system is $3^{18} \cdot 2^4 = 6\,198\,727\,824$. We do not encourage you to try to solve it.

Of course, this is not a clever way to compute the polynomials in question. However, the six billion degree of the system may be considered as a reasonable measure of difficulty of the problem. It is not without reason that it has taken 35 years between the enunciation of the problem and the computation of the above example. In fact, as we will see later (Section 9.5, page 146), for the degree 30 there are exactly four non-equivalent solutions (the definition of equivalence will be given later). All the other solutions are "parasitic" ones. Indeed, the system of equations of the above exercise does not ensure that the polynomials A and B will be coprime and that the difference $A^3 - B^2$ will be non-zero. We may of course add these conditions to the system but it will not make its resolution any easier. On the contrary, it will make it even more difficult to solve. It is also quite surprising that such a horrifying system has a solution in $\mathbb{Q}$. In general, the coefficients of the polynomials we are looking for are algebraic numbers: see the next section.

The following was conjectured in the above-cited paper [**BCHS-65**].

CONJECTURE 1.4 (Cube minus square). Let $A, B \in \mathbb{C}[x]$ be coprime and $\deg A = 2k$, $\deg B = 3k$, so that $\deg A^3 = \deg B^2 = 6k$. Then

(1) For $R = A^3 - B^2$ one always has $\deg R \geq k + 1$.
(2) This bound is sharp: that is, it is attained for infinitely many values of k.

In Example 1.2, we have had $k = 5$.

The first statement of the conjecture was proved the same year by Davenport [**Dav-65**]. The second one turned out to be much more difficult and remained open for 16 years: in 1981 Stothers [**Sto-81**] showed that the bound is in fact attained not only for infinitely many values of k but for all of them. We will see a spectacularly simple proof of this statement in Example 2.14, page 13.

The above problem may be generalized in various ways. The following one was considered in 1995 by Zannier [**Zan-95**]. Let $\alpha, \beta \vdash n$ be two partitions of n,

$$\alpha = (\alpha_1, \ldots, \alpha_p), \quad \beta = (\beta_1, \ldots, \beta_q), \quad \sum_{i=1}^{p} \alpha_i = \sum_{j=1}^{q} \beta_j = n,$$

and let P and Q be two coprime polynomials of degree n having the following factorization patterns:

$$P(x) = \prod_{i=1}^{p} (x - a_i)^{\alpha_i}, \qquad Q(x) = \prod_{j=1}^{q} (x - b_j)^{\beta_j}. \tag{1.4}$$

In these expressions we consider the root multiplicities α_i and β_j, $i = 1, 2, \ldots, p$, $j = 1, 2, \ldots, q$ as being given, while the roots a_i and b_j are not fixed, though they must all be distinct. The problem is to find the minimum possible degree of the difference $R = P - Q$. In his paper, Zannier proved the following statement.

THEOREM 1.5 (Zannier, 1995 [**Zan-95**]). *Let $d = \gcd(\alpha, \beta)$ be the greatest common divisor of the numbers $\alpha_1, \ldots, \alpha_p, \beta_1, \ldots, \beta_q$. If*

$$p + q \le \frac{n}{d} + 1 \tag{1.5}$$

then

$$\deg R \ge (n+1) - (p+q), \tag{1.6}$$

and this bound is always attained, whatever are α and β satisfying (1.5). *If, on the other hand, $p + q > \frac{n}{d} + 1$, then a weaker bound*

$$\deg R \ge \frac{(d-1)\,n}{d}, \tag{1.7}$$

is valid, and it is also attained.

REMARK 1.6 (Permutations vs. polynomials). In fact, a result equivalent to that of Zannier was obtained in 1982 by Boccara [**Boc-82**]. However, Boccara worked with permutations and was not aware of the fact that his considerations had something to do with polynomials. Subsequent authors also overlooked his paper. Profound relations of the above problem with permutation groups will be studied in Chapter 7.

EXAMPLE 1.7 (Difference of powers). Let $A, B \in \mathbb{C}[x]$ be coprime, $\deg A = kt$, $\deg B = ks$ where s and t are also coprrime and $s, t \ge 2$. Then we have $\deg A^s = \deg B^t = kst$, so that

$$n = kst, \quad \alpha = (\underbrace{s, s, \ldots, s}_{kt}) = s^{kt}, \quad \beta = (\underbrace{t, t, \ldots, t}_{ks}) = t^{ks}.$$

According to (1.4), we have $P = A^s$ where $A = \prod_{i=1}^{kt}(x - a_i)$, and $Q = B^t$ where $B = \prod_{j=1}^{ks}(x - b_j)$. Therefore, $p = kt$, $q = ks$, and

$$(n+1) - (p+q) = k(st - s - t) + 1.$$

For $s = 3$, $t = 2$ we obtain the previously conjectured bound $k + 1$.

REMARK 1.8 (Multiple roots). Notice that *a priori* polynomials A or B in the expression $A^s - B^t$ might have multiple roots. Suppose, for example, that the partition α contains some parts equal not to s but to $2s$, or $3s$, etc. In this case, the number p of the parts of α will be less that kt, and therefore, according to the above theorem, the lower bound $k(st - s - t) + 1$ will not be attained. However, in some specific situations the polynomials not attaining the bound may be interesting in their own right: see in this respect Examples 9.11 and 9.12, page 141 ff.

The above results justify our choice of terminology, attributing the names of Harald Davenport and Umberto Zannier to the polynomials in question[1].

DEFINITION 1.9 (Davenport–Zannier pair). Let $P, Q \in \mathbb{C}[x]$ be two coprime polynomials with factorization patterns (1.4), $\deg P = \deg Q = n$, while the degree of the polynomial $R = P - Q$ equals $(n+1)-(p+q)$. Then the pair (P, Q) is called *Davenport–Zannier pair*, or, in a more concise way, *DZ-pair*.

1.2. Dessins d'enfants and Galois theory

In general, the theory of dessins d'enfants studies meromorphic functions on *Riemann surfaces* having three critical values. Our part of affair is the most elementary one: instead of considering arbitrary Riemann surfaces we consider only the usual *Riemann complex sphere*. Thus, meromorphic functions become just ordinary rational functions. A rational function f is a *Belyĭ function* if its only critical values are 0, 1 and ∞. (A value is *critical* if its preimage contains multiple roots or, in the case of infinity, multiple poles.) Then, the preimage of the segment $[0, 1]$ under f is a map drawn on the sphere (see Remark 1.10). This map has a natural bicolored structure: we color the preimages of 0 in black, and the preimages of 1, in white. Inside each face there is a unique pole whose multiplicity is equal to the *degree of the face* (see Definition 1.11). Many properties of Belyĭ functions can be "seen" by looking at these maps.

Finally, if (P, Q) is a DZ-pair and $R = P - Q$ then the function $f = P/R$ is a Belyĭ function. In Section 1.3 we explain what is the class of the corresponding maps.

The problem of minimizing the degree of the difference $R = P - Q$ for P and Q like in (1.4), is *a priori* stated for polynomials with arbitrary complex coefficients. At the same time, Example 1.2 may give us an impression that it is always possible to find a solution with rational coefficients. However, this latter impression is too optimistic. What is true is that the coefficients of DZ-polynomials are algebraic numbers. More exactly: every DZ-pair *can be realized* over the field $\overline{\mathbb{Q}}$ of algebraic numbers. This situation leads us aside from our initial motivation (how close two powers of *integers* could be; a useful information concerning this question may be obtained only through the use of polynomials with rational coefficients) but it provides us with an incomparably more exciting one: the relation to Galois theory. Indeed, acting simultaneously on all the coefficients of a DZ-pair by the *absolute Galois group* $\mathrm{Gal}(\overline{\mathbb{Q}}|\mathbb{Q})$, which is the automorphism group of the field $\overline{\mathbb{Q}}$ of algebraic numbers, we get another DZ-pair. What do these pairs have in common? What are the invariants of this action? When do two DZ-pairs *not* belong to the same orbit of the action? All these, and many other questions will be studied in this book.

Still, the most important, and the most interesting case remains that of rational coefficients. Though now it becomes a particular case of the Galois action: namely, it is the case when the Galois orbit consists of a single element. An important step in this study was made in 2010 by Beukers and Stewart [**BeSt-10**]. They considered Davenport–Zannier pairs (which they call Davenport pairs) with rational coefficients having the form (A^s, B^t), like in Example 1.7. They found several infinite series of such

[1] In [**Zan-01**], Zannier also considers the same problem for polynomials over fields of finite characteristic. Some entirely new phenomena take place in this setting. However, this subject lies outside the scope of our book.

pairs, and also several sporadic examples. This is the general pattern of many classifications of dessins d'enfants. In what follows, we also draw a particular attention to the DZ-pairs with rational coefficients (in other terminology, *defined over* $\mathbb{Q}$).

The main notions and results of the theory of dessins d'enfants will be introduced in Chapter 2. Relations to Galois theory is certainly the most interesting aspect of this theory but there are also many other ramifications of the subject. In general terms, the possibility to "draw" functions provides us with many useful insights concerning the properties of the functions in question.

1.3. What are weighted trees

The notion of a weighted (bicolored plane) tree was defined in Preface (see p. viii). It was also mentioned there that, in fact, an edge of weight k represents a strand of k parallel edges: see Figure 1.1. Thus, weighted trees represent a specific class of plane maps. It turns out that technically it is much easier to work with trees than with maps, especially when the number of edges in a strand is not a constant but a variable.

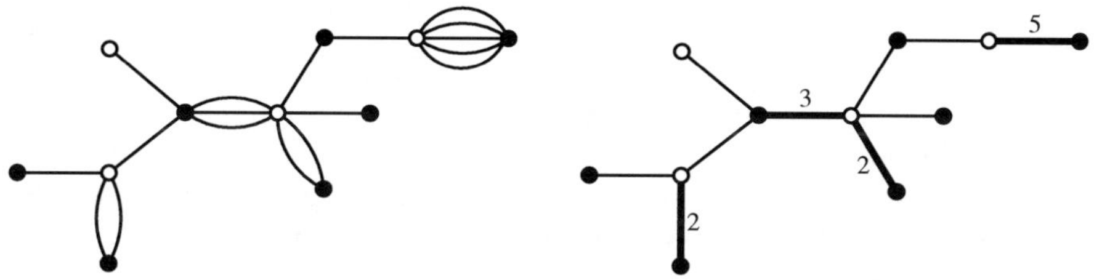

FIGURE 1.1. The passage from a map with parallel edges to the corresponding weighted tree.

REMARK 1.10 (On terminology). The term *map* is used in the literature in two different meanings. For example, the authors of the book [**BoSt-04**] and the authors of the paper [**SiVo-14**] use the word "map" as a synonym of the word *mapping*, that is, a function. They call *Belyĭ maps* what we call in this book *Belyĭ functions*. According to the second interpretation of the word "map", a map is a graph drawn on a two-dimensional surface in such a way that its edges do not intersect. This meaning is used, for example, in the context of the *map color theorem* (see [**Rin-74**]). A map is thus a subdivision of the corresponding surface into regions homeomorphic to an open disk; these regions are called *faces* of the map.

Throughout this book, we will use the word "map" in this second meaning.

We also introduce the notion of *degree of a face* which is specific for bicolored maps. Pay attention to this definition, it is not entirely standard.

DEFINITION 1.11 (Degree of a face). The *degree of a face* of a bicolored map is *the half* of the number of edges surrounding this face.

In this way, we have the sum of the degrees of the black vertices equal to the sum of the degrees of the white vertices, equal to the sum of the degrees of the faces and equal to the total weight of the tree. We will also use the term *degree of a tree* as a synonym of its total weight. Notice also that, according to the above definition, the degrees of all the faces of the map of Figure 1.1 except the outer face are equal to 1.

What is the relation between DZ-pairs and weighted trees? Let (P, Q) be a DZ-pair, and $R = P - Q$. Then the map like the one on the left in Figure 1.1 may be realized as a preimage of the segment $[0, 1]$ under the rational function $f = P/R$. In particular, the black vertices of this map are the roots of P, and the white vertices are the roots of Q, with the degrees of vertices being equal to the multiplicities of these roots. Thus, we may predict many properties of DZ-pairs by just looking at the corresponding pictures. Chapter 2 explains this construction in more detail. In fact, the (equivalence classes of) DZ-pairs are in a bijection with the (equivalence classes of) weighted bicolored plane trees.

EXAMPLE 1.12 (Computer plot). Let us consider the following DZ-pair:

$$\begin{aligned} P &= (x^5 + 50x^3 + 500x + 500)^9, \\ Q &= (x^9 + 90x^7 + 2700x^5 + 900x^4 + 30\,000x^3 + 36\,000x^2 \\ &\quad + 90\,000x + 180\,000)^5. \end{aligned}$$

The polynomial R here is of degree $(45 + 1) - (9 + 5) = 32$, and it is too cumbersome, so we do not write it explicitly. The pair (P, Q) corresponds to the weighted tree shown in Figure 1.2. This figure is a schematic representation of the tree in question while Figure 1.3 shows the "true form" of the same tree (or, rather, of the map encoded by this tree) as a preimage of the segment $[0, 1]$ under the rational function $f = P/R$. We did not mark vertices of this map in black and white since in this example it is clear: all vertices of degree 9 are black, and all vertices of degree 5 are white. We see that the schematic Figure 1.2 is, in a way, more informative. For example, it is difficult to see, in Figure 1.3, what takes place at the upper and lower ends of the tree. While the true form is maybe more beautiful[2], the schematic figure is more "readable": it is easier to see various characteristics of the tree in question. The reader is advised to compare the two pictures.

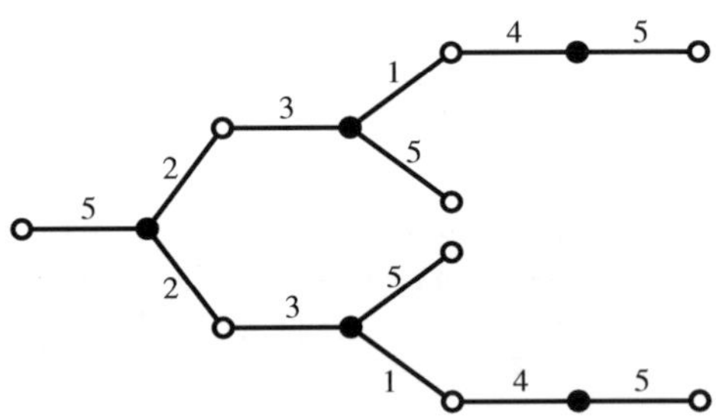

FIGURE 1.2. The weighted tree corresponding to the above DZ-pair of polynomials. The passport of this tree is $(9^5, 5^9)$.

In what follows, we will use exclusively the schematic pictures of the trees. As to this particular example, we will return to it twice: in Example 9.21 and in Example 11.17.

[2]It also provides us with a positive answer to an old question: is it possible for an edge of a dessin to contain an inflexion point?

FIGURE 1.3. A computer-made plot of the tree of Figure 1.2 realized as a preimage of the segment $[0, 1]$ under a Belyĭ function.

CHAPTER 2

Dessins d'enfants: from polynomials through Belyĭ functions to weighted trees

2.1. Rational function $f = P/R$ and its critical values

Let $\alpha, \beta \vdash n$ be two partitions of n, $\alpha = (\alpha_1, \dots, \alpha_p)$, $\beta = (\beta_1, \dots, \beta_q)$, $\sum_{i=1}^{p} \alpha_i = \sum_{j=1}^{q} \beta_j = n$, and let $P, Q \in \mathbb{C}[x]$ be two polynomials of degree n with the factorizations

$$P(x) = \prod_{i=1}^{p} (x - a_i)^{\alpha_i}, \qquad Q(x) = \prod_{j=1}^{q} (x - b_j)^{\beta_j}. \tag{2.1}$$

We suppose all $a_i, b_j, i = 1, \dots, p, j = 1, \dots, q$ to be distinct. Let the difference $R = P - Q$ have the following factorization:

$$R(x) = \prod_{k=1}^{r} (x - c_k)^{\gamma_k}, \qquad \deg R = \sum_{k=1}^{r} \gamma_k. \tag{2.2}$$

Our goal is to minimize $\deg R$; obviously,

$$\deg R \geq r. \tag{2.3}$$

Consider the following rational function of degree n:

$$f = \frac{P}{R};$$

note that

$$f - 1 = \frac{Q}{R}.$$

DEFINITION 2.1 (Critical value). A point $y \in \overline{\mathbb{C}} = \mathbb{C} \cup \{\infty\}$ is called *critical value* of a rational function f if the equation $f(x) = y$ has multiple roots in $\overline{\mathbb{C}}$.

The expressions written above for the function $f = P/Q$ provide us with at least three critical values of f:

- $y = 0$, provided that not all α_i are equal to 1;
- $y = 1$, provided that not all β_j are equal to 1; and
- $y = \infty$, if only we do not consider the trivial case $\deg R = \deg P - 1$; if $\deg R < \deg P - 1$ then f has a multiple pole at infinity.

Denote $y_1, \dots, y_m$ the other critical values of f, if there are any, and let n_l be the number of preimages of y_l, $l = 1, \dots, m$; by the definition of a critical value, $n_l < n$.

LEMMA 2.2 (Number of roots of R). *The number r of distinct roots of the polynomial R is*

$$r = (n + 1) - (p + q) + \sum_{l=1}^{m} (n - n_l). \tag{2.4}$$

In fact, equality (2.4) is a particular case of the Riemann–Hurwitz formula, but for the sake of completeness we give its proof here.

PROOF. Let us draw a *star-tree* with the center at 0 and with its rays going to the critical values $1, y_1, \ldots, y_m$, see Figure 2.1. Considered as a map on the sphere, this tree has $m+2$ vertices, $m+1$ edges, and a single outer face with its "center" at ∞.

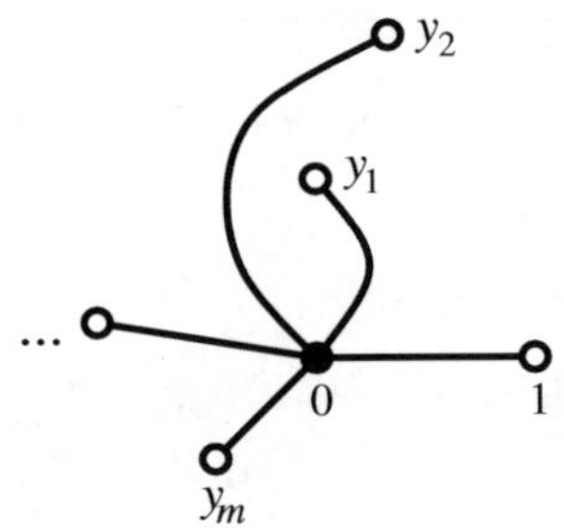

FIGURE 2.1. Star-tree whose vertices are critical values of f.

Now let us take the preimage of this tree under f; recall that $\deg f = n$. We will get a graph drawn on the preimage sphere which has $n(m+1)$ edges since each edge is "repeated" n times in the preimage. Its vertices are the preimages of the points $0, 1, y_1, \ldots, y_m$, so their number is equal to $p + q + \sum_{l=1}^{m} n_l$.

What occurs to the faces?

If we puncture the single open face in the image sphere at ∞, we get a punctured disk without any ramification points inside. The only possible unramified covering of a punctured open disk is a disjoint collection of punctured disks; their number is equal to the number of poles of f, namely, $r+1$ (r roots of R and ∞). Inserting a point into each puncture we get $r+1$ simply connected open faces in the preimage sphere. The fact that they are simply connected implies that the graph drawn on the preimage sphere is connected. Thus, the preimage of our star-tree is a plane map. What remains is to apply Euler's formula:

$$\left(p + q + \sum_{l=1}^{m} n_l\right) - n(m+1) + (r+1) = 2,$$

which leads to (2.4). □

Notice that in order to prove Lemma 2.2, instead of the tree of Figure 2.1, we could take any other plane map with vertices at the critical values (see, e. g., the proof of Proposition 3.8 below).

COROLLARY 2.3 (Lower bound). *We have*

$$\deg R \geq (n+1) - (p+q). \tag{2.5}$$

The proof follows from (2.4) and (2.3). □

Note that $\deg R$ cannot be negative; therefore, when $p+q > n+1$ the latter bound cannot be attained. In this case one can attain the bound $\deg R \geq 0$, that is, the polynomial R can be made equal to a constant. This situation is studied in more detail at the end of Chapter 3.

Equation (2.4) provides us with guidelines of how to get the minimum degree of R.

PROPOSITION 2.4 (Bound (2.5) attainability). *Bound* (2.5) *is attained if and only if the following conditions are satisfied:*

- $p+q \leq n+1$.
- *The number m of the critical values of f other than* $0, 1, \infty$, *is equal to zero, so that the sum* $\sum_{l=1}^{m}(n-n_l)$ *in the right-hand side of* (2.4) *is altogether eliminated. The tree of* Figure 2.1 *is then reduced to merely the segment* $[0,1]$.
- *All the roots of R are simple, that is,* $\gamma_1 = \ldots = \gamma_r = 1$, *so that* $\deg R = r$. *Another formulation of the same condition is as follows: the partition* $\gamma \vdash n$, $\gamma = (\gamma_0, \gamma_1, \gamma_2, \ldots, \gamma_r)$, *which corresponds to the multiplicities of the poles, has the form of a hook:* $\gamma = (n-r, \underbrace{1, 1, \ldots, 1}_{r \text{ times}}) = (n-r, 1^r)$.

The conditions which imply the existence of such a function f will be obtained in Chapter 3.

2.2. Dessins d'enfants and Belyĭ functions

Considering rational functions with only three critical values brings us into the realm of the theory of *dessins d'enfants*. Here we give a brief summary of this theory (only in a planar setting); the missing details, proofs, and bibliography can be found, for example, in [**JoWo-16**], [**GiGo-12**] or in Chapter 2 of [**LaZv-04**].

DEFINITION 2.5 (Belyĭ function). A rational function $f : \overline{\mathbb{C}} \to \overline{\mathbb{C}}$ is called *Belyĭ function* if it does not have critical values outside the set $\{0, 1, \infty\}$.

Let f be such a function. Then the tree considered in the proof of Lemma 2.2 is reduced to the segment $[0,1]$. Let us take this segment, color the point 0 in black and the point 1 in white, and consider the preimage $D = f^{-1}([0,1])$; we will call this preimage a *dessin*.

PROPOSITION 2.6 (Dessin). *The dessin* $D = f^{-1}([0,1])$ *is a connected graph drawn on the sphere, and its edges do not intersect outside the vertices. Therefore, D may also be considered as a plane map. This map has a bipartite structure: black vertices are preimages of* 0, *white vertices are preimages of* 1, *and the ends of each edge have the opposite colors.* □

The degrees of the black vertices are equal to the multiplicities of the roots of the equation $f(x) = 0$, and the degrees of the white ones are equal to the multiplicities of the roots of the equation $f(x) = 1$. The sum of the degrees in both cases is equal to $n = \deg f$, which is also the number of edges.

The map D being bipartite, the number of edges surrounding each face is even. It is convenient, in defining the face degrees, to divide this number by two.

DEFINITION 2.7 (Face degree; incidence of edges and faces). Suppose that we are inside of face and make a circuit of it near its boundary in the positive trigonometric direction. We say that *an edge is incident to the face* if we follow this edge from its black end toward the white one. Thus, only half of the edges surrounding a face are incident to it. Moreover, each edge is incident to exactly one face. The *degree of a face* is equal to the number of edges incident to it.

REMARK 2.8 (Outer face). When we go around the outer face, the trigonometric direction should be considered from the observation point on the opposite side of the sphere. From "our" point of view in front of the sphere the same circuit looks as if it turns in the negative trigonometric direction. See also Convention 7.2 (page 88), Example 7.3 and Figure 7.1.

According to the above definition and to the remarks preceding it, every edge is incident to one black vertex, to one white vertex, and to one face. The sum of the face degrees is thus equal to $n = \deg f$.

PROPOSITION 2.9 (Faces and poles). *Inside each face there is a single pole of f, and the multiplicity of this pole is equal to the degree of the face.* □

DEFINITION 2.10 (Passport of a dessin). The triple $\pi = (\alpha, \beta, \gamma)$ of partitions α, β, $\gamma \vdash n$ which correspond to the degrees of the black vertices, of the white vertices, and of the faces of a dessin, is called a *passport* of the dessin.

DEFINITION 2.11 (Combinatorial orbit). A set of the dessins having the same passport is called a *combinatorial orbit* corresponding to this passport.

Now comes a very important moment which is crucial for the whole book:

The construction which associates a map to a Belyĭ function works also in the opposite direction.

Two bicolored plane maps are *isomorphic* if there exists an orientation preserving homeomorphism of the sphere which transforms one map into the other, respecting the colors of the vertices. Let M be a bicolored map on the sphere. Then, the sphere may be endowed with a complex structure, thus becoming the Riemann complex sphere, and a representative of the isomorphism class of M can be drawn as a dessin D obtained via a Belyĭ function. The following statement is a particular case of the classical Riemann's existence theorem:

PROPOSITION 2.12 (Existence of Belyĭ functions). *For every bicolored plane map M there exists a dessin $D = f^{-1}([0,1])$ isomorphic to M, where f is a Belyĭ function. This function $f = f(x)$ is unique up to a linear fractional transformation of the variable x.* □

Of course, in this book, when we draw a map we do not respect the specific geometric form of the corresponding dessin like it was the case in Figure 1.3. We are content with the fact that such a dessin does exist.

Now Proposition 2.4 may be reformulated in purely combinatorial terms:

PROPOSITION 2.13 (Bound (2.5) attainability). *The lower bound* (2.5) *is attained if and only if there exists a bicolored plane map with the passport $\pi = (\alpha, \beta, \gamma)$ in which the partitions $\alpha = (\alpha_1, \ldots, \alpha_p)$ and $\beta = (\beta_1, \ldots, \beta_q)$ are given, and γ has the form $\gamma = (n - r, 1^r)$ where 1 is repeated $r = (n+1) - (p+q)$ times.* □

In geometric terms, all the faces of our map except the outer one must be of degree 1. Recall that the *number* of faces, which is equal to $r + 1$, is prescribed by Euler's formula.

EXAMPLE 2.14 (Cubes and squares: a solution). Let us look once again at the problem posed by Birch et al. in **[BCHS-65]** (see page 2). In order to show that if $\deg A = 2k$, $\deg B = 3k$, and $R = A^3 - B^2$, then the lower bound $\deg R \geq k + 1$ can be attained, we must construct a map with the following properties: all its black vertices are of degree 3; all its white vertices are of degree 2; and all its finite faces are of degree 1.

In order to simplify our pictures we sometimes use the following convention.

CONVENTION 2.15 (Three ways of drawing maps). When all the white vertices are of degree 2, it is convenient, in order to simplify a graphical representation of the corresponding maps, to draw only black vertices and to omit the white ones, considering them as being implicit. In such a picture, a line connecting two black vertices contains an invisible white vertex in its middle, and thus is not an edge but a union of two edges.

Figure 2.2 shows three ways of drawing the same map. The loop in the leftmost picture represents a face of degree 1, and this is coherent with our Definition 2.7.

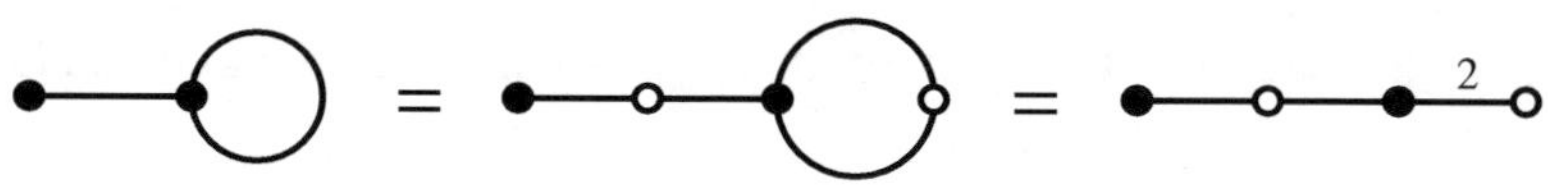

FIGURE 2.2. These three pictures represent the same map.

Now the construction of maps we need in order to solve the above problem about $\min \deg(A^3 - B^2)$ becomes trivial: first we draw a tree with all internal nodes being of degree 3, and then attach loops to its leaves: see Figure 2.3. □

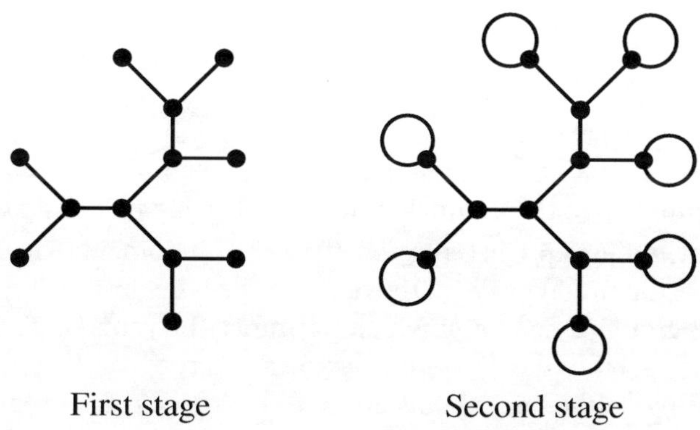

FIGURE 2.3. This map solves the problem which remained open for 16 years: the mere existence of a picture like the one on the right implies the existence of polynomials A and B, $\deg A = 2k$, $\deg B = 3k$, such that $\deg(A^3 - B^2) = k + 1$.

We see in this example a remarkable efficiency of the pictorial representation of problems concerning polynomials. If this representation was known in 1965, the proof of the conjecture would have taken 16 minutes instead of 16 years.

2.3. Number fields

As it was stated in Proposition 2.12, a Belyĭ function $f(x)$ corresponding to a dessin is defined up to a linear fractional transformation of the variable x. In this family of equivalent Belyĭ functions it is always possible to find one whose coefficients are algebraic numbers. If we act simultaneously on all the coefficients of such a function by an element of the *absolute Galois group* $\mathrm{Gal}\,(\overline{\mathbb{Q}}|\mathbb{Q}) = \mathrm{Aut}(\overline{\mathbb{Q}})$, that is, by an automorphism σ of the field $\overline{\mathbb{Q}}$ of algebraic numbers, or, in other words, if we replace all the coefficients a_i of f by their algebraically conjugate numbers $\sigma(a_i)$, we obtain once again a Belyĭ function. Furthermore, one can prove that in such a way the action of $\mathrm{Gal}\,(\overline{\mathbb{Q}}|\mathbb{Q})$ on Belyĭ functions descends to an action on dessins. There exist many combinatorial invariants of this action, the first and the simplest of them being the passport of the dessin. Thus, a combinatorial orbit (see Definition 2.11) may constitute a single Galois orbit, or may further split into a union of several Galois orbits. Every combinatorial orbit is finite, and therefore every Galois orbit is also finite.

Number field is a finite algebraic extension of the field $\mathbb{Q}$. Three different but closely related constructions of number fields are used when we study the Galois action on dessins d'enfants. The most important of these constructions is that of the field of moduli.

CONSTRUCTION 2.16 (Field of moduli). Let D be a dessin, and consider the subgroup $\Gamma_D \leq \mathrm{Gal}\,(\overline{\mathbb{Q}}|\mathbb{Q})$ which is the stabilizer of D. Since the orbit of D is finite, the group Γ_D is a subgroup of finite index in $\mathrm{Gal}\,(\overline{\mathbb{Q}}|\mathbb{Q})$. Let $H \leq \Gamma_D$ be the maximal normal subgroup of $\mathrm{Gal}\,(\overline{\mathbb{Q}}|\mathbb{Q})$ contained in Γ_D. According to the Galois correspondence between subgroups of $\mathrm{Gal}\,(\overline{\mathbb{Q}}|\mathbb{Q})$ and algebraic extensions of $\mathbb{Q}$, there exists a number field K corresponding to the group H. This field is called the *field of moduli* of the dessin D. By construction, this field is unique: a dessin cannot have two different fields of moduli.

Let $\mathcal{D} = \{D_1, \ldots, D_m\}$ be an orbit of the Galois action on dessins. Then there exists a polynomial $T \in \mathbb{Q}[t]$ of degree m, irreducible over $\mathbb{Q}$, such that the field of moduli K is the *splitting field* of T. The latter means that K is generated by the roots of T. We call T *defining polynomial* of K. Unlike the field itself, its defining polynomial is not at all unique: determining if two polynomials generate the same field is a task in itself.

The action of the group $\mathrm{Gal}\,(\overline{\mathbb{Q}}|\mathbb{Q})$ on the orbit $\mathcal{D}$ coincides with the action of $\mathrm{Gal}\,(K|\mathbb{Q})$ which is the *Galois group* of the field K. The latter group may be represented as a permutation group of degree m which permutes the roots of T. It is also called Galois group of the polynomial T. We may associate m roots of the polynomial T to m dessins of the orbit $\mathcal{D}$. When the Galois group sends a root t_i to another root t_j, it also sends the dessin D_i to the dessin D_j.

In the absolute majority of cases the situation looks as follows. The coefficients of a Belyĭ function corresponding to a dessin D_i are expressed in terms of the corresponding root t_i. Replacing simultaneously t_i with t_j in all these expressions produces a Belyĭ function for the dessin D_j. However, in some specially constructed examples one needs a larger field $L \supset K$ to be able to find a corresponding Belyĭ function. The field L is called *field of definition* or *field of realization* of the dessin. It is not unique: various fields may serve as fields of realization of the same dessin. The action of $\mathrm{Gal}\,(L|K)$ on Belyĭ functions may change a position of a dessin $D \in \mathcal{D}$ on the complex sphere but does not change its combinatorial structure; in other words, as a map, the dessin in question remains the same.

There is a simple *sufficient condition* which ensures that that there exists a Belyĭ function with the coefficients in the field of moduli, see [**Cou-94**] or [**SiVo-16**]: this condition is the existence of a *bachelor*.

DEFINITION 2.17 (Bachelor). A *bachelor* is a black vertex (respectively, a white vertex or a face) such that there are no other black vertices (respectively, no other white vertices or no other faces) of the same degree.

REMARK 2.18 (Positioning of bachelors). If a dessin contains several bachelors then up to three of them can be placed at rational points, that is, at points in $\mathbb{Q} \cup \{\infty\}$, and this will not prevent the Belyĭ function of the dessin in question to be defined over the field of moduli.

For the dessins we study in this book a bachelor always exists: it is the outer face (since all the other faces are of degree 1). Naturally, we place its center at ∞. Thus, *in our setting the field of realization always coincides with the field of moduli.*

The defining polynomial T contains all the information one needs to know about the Galois action on the orbit $\mathcal{D}$, but some parts of this information may be difficult to extract. For example, the *degree of a number field* is defined as the dimension of this field as a vector space over $\mathbb{Q}$. This dimension is equal to the order of the Galois group of this field, or the Galois group of its defining polynomial. Thus, the degree of the field of moduli K is equal to the order of the Galois group of the polynomial T and may therefore vary between m and $m!$, where $m = \deg T$. However, this group and its order may be very difficult to find. Usually this does not create any problem since we rarely need to know these particular parameters. As a rule, we work not with the field K itself but with a smaller field $\mathbb{Q}(t_i)$ generated not by all the roots of T but by one of them (though an arbitrary one). All these fields $\mathbb{Q}(t_i)$, for $i = 1, \ldots, m$, are isomorphic to each other, and they are isomorphic to the field of residues modulo T in $\mathbb{Q}[t]$. They constitute, in fact, different embeddings of the latter field in $\overline{\mathbb{Q}}$. They are extensions of degree m of $\mathbb{Q}$. Obviously, the field of residues modulo T is easy to manipulate. Quite often, by abuse of terminology, people say that the splitting field of T is an extension of $\mathbb{Q}$ of degree m.

Sometimes, when the roots of T admit an explicit representations, we, by abuse of notation, take a particular root t_i and denote the field of moduli by $\mathbb{Q}(t_i)$. We do that since it is usually easier to find T knowing t_i than otherwise.

EXAMPLE 2.19 (Biquadratic polynomial). Let us illustrate the above remarks by an example. Consider the biquadratic polynomial $T = t^4 + pt^2 + q$. Its four roots are

$$\pm\sqrt{\frac{-p \pm \sqrt{p^2 - 4q}}{2}}.$$

It is well known that the Galois group G of T is determined as follows:

- When q is a perfect square then G is Klein's four-group, $G = V_4 = C_2 \times C_2$, $|G| = 4$.
- When q is not a perfect square but $q(p^2 - 4q)$ is, then G is the cyclic group, $G = C_4$, $|G| = 4$.

- When neither q nor $q(p^2-4q)$ is a perfect square then G is the dihedral group, $G = \mathrm{D}_4$, $|G| = 8$.
- If both q and $q(p^2-4q)$ are perfect squares the polynomial T is reducible, so that the question about its Galois group makes no sense.

Let us take $T = t^4 - 46t^2 + 621$: it is one of the defining polynomials of the moduli field of the orbit 23.1 on Page 127. According to the above, its Galois group is D_4 of order 8. Its roots are

$$\pm\sqrt{23 \pm 2\sqrt{-23}}.$$

Multiplying two of these roots we get

$$\sqrt{23 + 2\sqrt{-23}} \cdot \sqrt{23 - 2\sqrt{-23}} = \sqrt{23^2 - 4 \cdot (-23)} = \sqrt{23 \cdot 27} = 3\sqrt{69}.$$

It is easy to verify that $\sqrt{69}$ does *not* belong to the field generated by one of the roots of T. To do that, we may write $r(t) = a_0 + a_1t + a_2t^2 + a_3t^3$, compute $r(t)^2 - 69 \bmod T$, equate the coefficients of the polynomial thus obtained to zero, solve the system (it is of degree 16) and see that the system does not have rational solutions. Or, otherwise, we may ask if $\sqrt{69}$ belongs to the field in question directly to the system Pari/GP (and get a negative answer). However, $\sqrt{69}$ obviously belongs to the field generated by all the four roots.

The similar observations can be made about the orbit 9.8, page 111. By contrast to these two cases, the Galois group of the orbit 8.1, page 104, and that of the orbit 8.6, page 105, is Klein's group V_4 of order 4, so that the moduli field both times is an extension of $\mathbb{Q}$ of degree 4. The same thing is true also for the orbit 13.1, page 119, though the defining polynomial is not biquadratic: this conclusion is based on the fact that the Galois group for this orbit is the cyclic group C_4 of order 4.

ORBITS OF SIZE ONE. Two chapters of our book are dedicated to the situation when a combinatorial orbit consists of a single element. In this case the Galois orbit also consists of a single element and therefore the corresponding number field is an "extension of degree one" of $\mathbb{Q}$, that is, it is the field $\mathbb{Q}$ itself.

PROPOSITION 2.20 (Coefficients in $\mathbb{Q}$). *If for a given passport $\pi = (\alpha, \beta, \gamma)$, where the partition γ is of the form $\gamma = (n-r, 1^r)$, there exists a unique bicolored plane map, then there exists a corresponding Belyĭ function with rational coefficients, and therefore there also exists a DZ-pair with rational coefficients.* □

Note that Proposition 2.13, which concerns the existence, is of the "if and only if" type, while Proposition 2.20, which concerns the definability over $\mathbb{Q}$, provides only an "if"-type condition.

In Chapter 5 we classify all the trees which are uniquely determined by their passport; such a tree constitutes an orbit all by itself. There are ten infinite series of such trees and ten sporadic trees which do not belong to any series. In Chapter 6 we compute all the corresponding Belyĭ functions and DZ-pairs.

2.4. Additional remarks about weighted trees

Now it becomes clear how the weighted trees appear in the study of DZ-polynomials. Combining together Propositions 2.4 and 2.13 we may state the following theorem:

THEOREM 2.21 (Lower bound attainability). *Let $\alpha, \beta \vdash n$ be two partitions of n having p and q parts, respectively. Then the lower bound (2.5) is attained if and only if there exists a weighted tree with the passport (α, β).* □

A simple condition which guarantees the existence of such a tree will be given in the next chapter. Here we add a few definitions.

DEFINITION 2.22 (Characteristics of weighted trees). The *weight distribution* of a weighted tree is a partition $\mu \vdash n$, $\mu = (\mu_1, \mu_2, \dots, \mu_m)$ where m is the number of edges, and μ_i, $i = 1, \dots, m$ are the weights of the edges. We have $m = p + q - 1$ where p is the number of black vertices and q is the number of white vertices.

Leaving aside the weights and considering only the underlying plane tree, we speak of a *topological tree*.

Weighted trees whose weight distribution is $\mu = 1^n$ will be called *ordinary trees*. Ordinary trees correspond to *Shabat polynomials*: these are particular cases of Belyĭ functions, with a single pole at infinity.

We call a *leaf* a vertex which has only one edge incident to it, whatever is the weight of this edge. By abuse of language, we will also call a leaf this edge itself.

One should not confuse the meaning of the adjectives *planar* and *plane*. A *graph* is planar if it is possible to draw it on the plane or on the sphere in such a way that its edges do not intersect. Obviously, any tree is planar. A *map* is plane if it is already drawn on the plane or on the sphere in a particular way, and its edges do not intersect. The trees of the above definition are considered not as mere graphs but as plane maps. More precisely, this means that the cyclic order of branches around each vertex of the tree is fixed. Changing this order will give the same tree considered as a graph but in general will give a different plane tree. *All the trees considered in this book will be endowed with the "plane" structure*; therefore, the adjective "plane" will often be omitted.

DEFINITION 2.23 (Isomorphic trees). Two weighted trees are *isomorphic* if the underlying bicolored plane maps are isomorphic. In other words, they are isomorphic if there exists a color-preserving bijection between the vertices of the trees and a bijection between the edges which respect the incidence of edges and vertices, the cyclic order of the edges around each vertex, and which also respect the weights of the edges.

DEFINITION 2.24 (Passport of a tree). The pair (α, β) of partitions $\alpha, \beta \vdash n$ of the total weight n of a tree, corresponding to the degrees of the black vertices and of the white vertices, is called the *passport* of this tree.

EXAMPLE 2.25 (Tree of Figure 1.1). The total weight of the tree shown in Figure 1.1 (page 5) is $n = 18$; its passport is $(\alpha, \beta) = (5^2 2^3 1^2, 7^1 6^1 4^1 1^1)$; the face degree distribution is $\gamma = 10^1 1^8$, and the weight distribution is $\mu = 5^1 3^1 2^2 1^6$.

REMARK 2.26 (Weighted trees vs. "weighted maps"). Albeit weighted trees are themselves maps, we must not confuse the notions of weighted trees and of "weighted maps". The weighted tree on the left, and the "weighted map" on the right of Figure 2.4 have the same set of black and white vertex degrees: $(\alpha, \beta) = (5^1 3^1 2^1, 5^2)$. However, the face degree partitions of the underlying maps are different: $\gamma = 4^1 1^6$ for the map represented by the tree, and $\gamma = 3^1 2^1 1^5$ for the map on the right. This latter map contains a face which is not the outer one, and which is not of degree 1: it is of degree 2.

Therefore, this map does not fit our construction. In particular, we cannot represent it by a weighted tree, and the two corresponding dessins cannot belong to the same combinatorial orbit and therefore to the same Galois orbit.

Through the whole book, we speak exclusively about weighted *trees*.

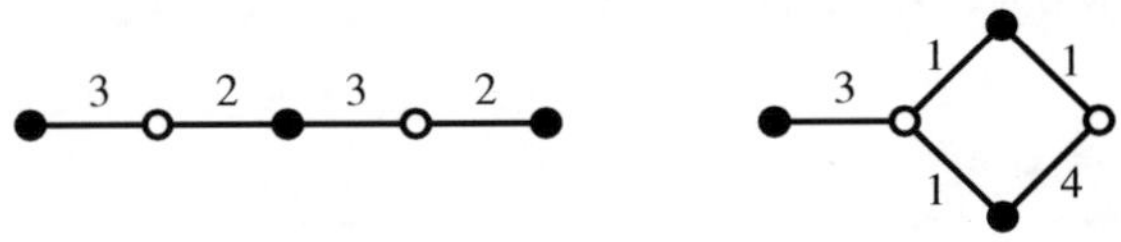

FIGURE 2.4. The weighted tree on the left and the "weighted map" on the right have the same set of black and white vertex degrees, but their face degrees are different. In particular, the map on the right has a face of a forbidden degree: it is not the outer face, and its degree is not equal to 1 (it is equal to 2). Therefore, this map cannot be represented by a weighted tree.

CHAPTER 3

Existence theorem

3.1. Realizability of coverings

In this chapter we study the following question: for a given pair of partitions $\alpha, \beta \vdash n$, does there exist a weighted tree of the total weight n with the passport (α, β)? Equivalently, does there exist a rational function with three critical values, and with the multiplicities of the preimages of these critical values being, first, two given partitions $\alpha, \beta \vdash n$, and then, the third partition being equal to $\gamma = (n-r, 1^r)$?

This question is a particular case of a more general problem of *realizability* of ramified coverings: does there exist a ramified covering of a given Riemann surface with the given "local data" (that is, with given multiplicities of the preimages of ramification points)? The problem goes back to the classical paper by Hurwitz [**Hur-1891**]. Though many particular cases are well studied, the problem in its full generality remains unsolved. Among numerous publications dedicated to the realizability we would like to mention early works by Husemoller [**Hus-62**] and Thom [**Tho-65**]; an important paper by Edmonds, Kulkarni and Stong [**EKS-84**]; and recent publications [**PaPe-09**], [**CPZ-12**], and [**Pak-09**].

In order to give an idea of the difficulty of the problem, we mention the following theorem which is not related to the subject of our book but which shows what could be possible types of answers to the question. Let us take $n = 4k + 1$, $k \geq 0$, and consider the following passport of a putative bicolored map with n edges: $(\alpha, \beta, \gamma) = (4^k 1^1, 2^{2k} 1^1, 4^k 1^1)$. If such a map exists, it is planar. Indeed, the number of vertices (black and white together) is $(k+1)+(2k+1) = 3k+2$; the number of edges is $4k+1$; the number of faces is $k + 1$. Thus, the Euler characteristic is equal to 2. Now, the following holds (see [**PaPe-09**] and [**CPZ-12**]).

THEOREM 3.1 (Sum of two squares). *A plane map with the passport* $(\alpha, \beta, \gamma) = (4^k 1^1, 2^{2k} 1^1, 4^k 1^1)$ *exists if and only if the number* $n = 4k + 1$ *is either a square,* $n = a^2$, *or a sum of two squares:* $n = a^2 + b^2$, $a, b \in \mathbb{Z}_+$. □

CONJECTURE 3.2 (Prime number of edges, [**EKS-84**]). If n is prime, the answer to the realizability question is always positive: a bicolored map does exist whatever is the passport of degree n satisfying the planarity condition.

The situation we work with in this book is much simpler. The main result of this chapter is the following theorem (recall that $\gcd(\alpha, \beta)$ denotes the greatest common divisor of the numbers $\alpha_1, \ldots, \alpha_p, \beta_1, \ldots, \beta_q$):

THEOREM 3.3 (Realizability of a passport by a tree). *Let* $\alpha, \beta \vdash n$ *be two partitions of* n, $\alpha = (\alpha_1, \ldots, \alpha_p)$, $\beta = (\beta_1, \ldots, \beta_q)$, *and let* $\gcd(\alpha, \beta) = d$. *Then a weighted tree with the passport* (α, β) *exists if and only if*

$$p+q \leq \frac{n}{d}+1. \tag{3.1}$$

The number of edges of the topological tree in question should be $m = p+q-1$. *Therefore, the above condition may be rewritten in the following simple way:*

$$dm \leq n. \tag{3.2}$$

By Theorem 2.21 the attainability of the bound (1.6) (coinciding with (2.5)) follows from this statement. The attainabilty of the bound (1.7) in the case when condition (3.1) is not satisfied will be established at the end of this chapter.

Theorem 3.3, and Theorem 3.7 below are equivalent to the main result of the paper [**Zan-95**] by Zannier (see Theorem 1 therein). In his paper, Zannier remarks that it would be interesting to apply the theory of dessins d'enfants to this problem in a more direct way, and mentions a remark made by Gareth Jones that such an approach might produce a simpler proof. This is indeed the case, as we will see in this section. Furthermore, this theory enables us to describe a huge class of DZ-pairs over $\mathbb{Q}$ (in a way, "almost all" of them), see Chapter 5; and it also gives us a more direct access to Galois theory, see Chapter 9. We have already had a first glimpse of the power of the "dessin method" in Example 2.14 (page 13).

3.2. Forests

A *forest* is a disjoint union of trees.

PROPOSITION 3.4 (Realizability of a passport by a forest). *Any pair* (α, β) *of partitions of* n *can be realized as a passport of a forest of weighted trees.*

PROOF. If there are two equal parts $\alpha_i = \beta_j$ in the partitions α and β, we create a separate edge with the weight $s = \alpha_i = \beta_j$ and proceed with the new passport (α', β'), where α' and β' are obtained from α and β by eliminating their parts α_i and β_j, respectively.

If there are no equal parts, suppose, without loss of generality, that there are two parts $\alpha_i > \beta_j$. Then we do the following (see Figure 3.1):

(a) make an edge with the weight $s = \beta_j$;
(b) consider the new passport (α', β') where β' is obtained from β by eliminating the part β_j, and α' is obtained from α by replacing α_i with $t = \alpha_i - \beta_j$;
(c) construct inductively a forest $\mathcal{F}'$ of total weight $n - s$ corresponding to the passport (α', β'); by definition, this forest must have a black vertex of degree t;
(d) glue the edge of weight s to the forest $\mathcal{F}'$ by fusing two vertices, as is shown in Figure 3.1, and get a forest $\mathcal{F}$ corresponding to (α, β) (since $s + t = \alpha_i$).

The proposition is proved. □

3.3. Stitching several trees to get one: the case $\gcd(\alpha, \beta) = 1$

THEOREM 3.5 (Existence). *Suppose that* $\gcd(\alpha, \beta) = 1$. *Then the passport* (α, β) *can be realized by a weighted tree if and only if* $p + q \leq n + 1$, *where* p *(respectively,* q*) is the number of parts in* α *(respectively, in* β*).*

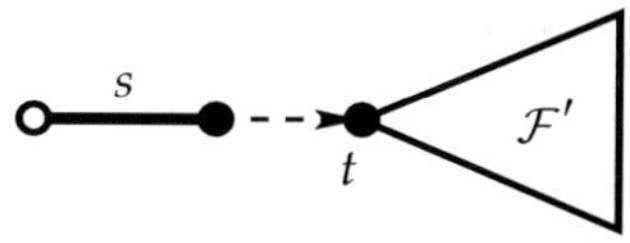

FIGURE 3.1. Construction of a forest in the case $\alpha_i > \beta_j$. Here $s = \beta_j$ and $t = \alpha_i - \beta_j$.

PROOF. According to Proposition 3.4 we may suppose that we already have a forest corresponding to the passport (α, β). Now suppose that there are two edges of weights s and u, $s < u$, which belong to different trees. Then we may stitch them together by the operation shown in Figure 3.2. The degrees of the vertices in the new, connected figure are the same as in the old, disconnected one. Figure 3.3 shows that the operation works in the same way when there are subtrees attached to the ends of the adjoined edges.

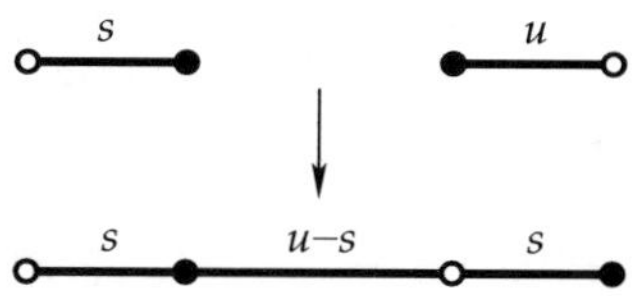

FIGURE 3.2. Stitching two edges.

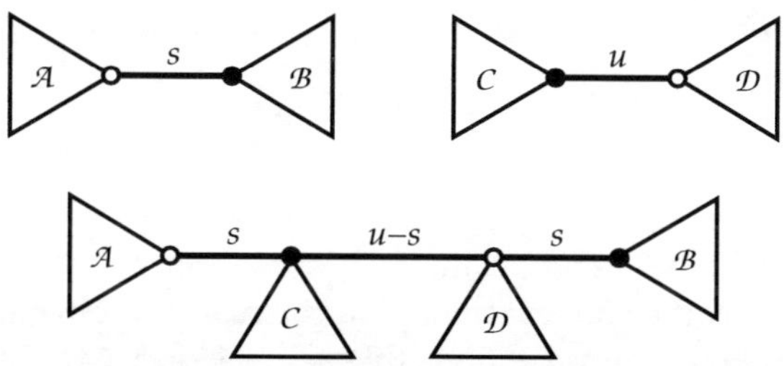

FIGURE 3.3. Stitching two trees.

We repeat this stitching operation until it becomes impossible to continue. The latter may happen in two ways. Either we have got a connected tree, and then we are done. Or there are no more edges with different weights while the forest remains disconnected. Then, taking into account that $\gcd(\alpha, \beta) = 1$, we conclude that the weights of all edges are equal to 1, that is, we have got a forest consisting of $l > 1$ ordinary trees. In this case, the number of vertices $p + q$ equals $n + l$ and therefore is strictly greater than $n + 1$, which contradicts the condition of the theorem. □

Note that the side edges in Figures 3.2 and 3.3 have the same weight s. We will use this property while performing the operation inverse to stitching, namely, the ripping of a connected tree in two, in the proof of Proposition 5.14 (see Figure 5.15, page 37).

3.4. Non-coprime weights

Now suppose that $\gcd(\alpha, \beta) = d > 1$.

LEMMA 3.6 (All the weights are multiples of $d > 1$). *The degrees of all vertices of a forest are divisible by $d > 1$ if and only if the weights of all edges are also divisible by d.*

PROOF. In one direction this is evident: the degrees of the vertices are sums of weights, and therefore, if all the weights are multiples of d, then the same is true for the degrees.

In the opposite direction, if all the vertex degrees are divisible by d, then it is true, in particular, for the leaves. Cut any leaf off the tree to which it belongs, and the statement is reduced to the same one for a smaller forest. □

Thus, dividing by d all the vertex degrees, that is, all the elements of the partitions α and β, we return to the situation of Theorem 3.5, with the same numbers p and q, and with the total weight equal to n/d. This finishes the proof of Theorem 3.3. □

We hope the reader will appreciate the simplicity of the above proof: number theorists have been approaching this result for 30 years (1965–1995). Once again, the credit goes to the pictorial representation of polynomials with the desired properties.

3.5. Weak bound: polynomials and cacti

When condition (3.1) of Theorem 3.3 is satisfied then, according to Theorem 2.21, the main bound (1.6) is attained. If this condition is *not satisfied*, then the following holds:

THEOREM 3.7 (Weak bound). *Let* $\gcd(\alpha, \beta) = d$, *and let* $p + q > \dfrac{n}{d} + 1$. *Then*

$$\deg R \geq \frac{(d-1)\,n}{d}, \tag{3.3}$$

and this bound is attained.

We will need the following proposition which was proved by Thom [**Tho-65**], and then reproved in many other publications. For the reader's convenience we provide a short proof based on the dessins d'enfants theory, see [**LaZv-04**], Corollary 1.6.9.

PROPOSITION 3.8 (Realizability of polynomials). *Let* $\Lambda = (\lambda_1, \lambda_2, \ldots, \lambda_k)$ *be a set of* $k \geq 1$ *partitions* $\lambda_i \vdash n$ *of number* n. *Denote by* p_i *the number of parts of* λ_i, $i = 1, 2, \ldots, k$. *Let* $y_1, y_2, \ldots, y_k \in \mathbb{C}$ *be arbitrary complex numbers. Then a necessary and sufficient condition for the existence of a polynomial* $T \in \mathbb{C}[x]$ *of degree* n, *all of whose finite critical values are contained in the set* $\{y_1, y_2, \ldots, y_k\}$, *and such that the multiplicities of the roots of the equation* $T(x) = y_i$ *correspond to the partitions* λ_i, $i = 1, 2, \ldots, k$, *is the following equality:*

$$\sum_{i=1}^{k} p_i = (k-1)\,n + 1\,, \quad \textit{or, equivalently,} \quad \sum_{i=1}^{k} (n - p_i) = n - 1\,. \tag{3.4}$$

PROOF. For purely aesthetic reasons, instead of taking a tree with the vertices $y_1, y_2, \ldots, y_k$, as we did in the proof of Lemma 2.2, let us take a Jordan curve J on the y-plane passing through the points $y_1, y_2, \ldots, y_k$, and let C be its preimage, $C = T^{-1}(J)$.

Then C is a tree-like map often called *cactus*: it does not contain any cycles except n copies of J glued together at the vertices, and these vertices are preimages of the points y_i. The number of copies of J glued together at a vertex is equal to the multiplicity of the corresponding root of the equation $T(x) = y_i$; see Figure 3.4. Equation (3.4) may then be interpreted as Euler's formula for the cactus since the cactus has $\sum_{i=1}^{k} p_i$ vertices, kn edges, and $n + 1$ faces (n copies of J and the outer face).

We leave it to the reader to verify that another proof of the necessity of formulas (3.4) can be deduced from the fact that the sum $\sum_{i=1}^{k}(n - p_i)$ in the second equality in (3.4) represents the degree of the derivative $T'(x)$.

These observations prove that conditions (3.4) are necessary. Notice that the partition $\lambda = 1^n$ may be eliminated from Λ (if it belongs to it), and may also be added to it, and this does not invalidate equalities (3.4).

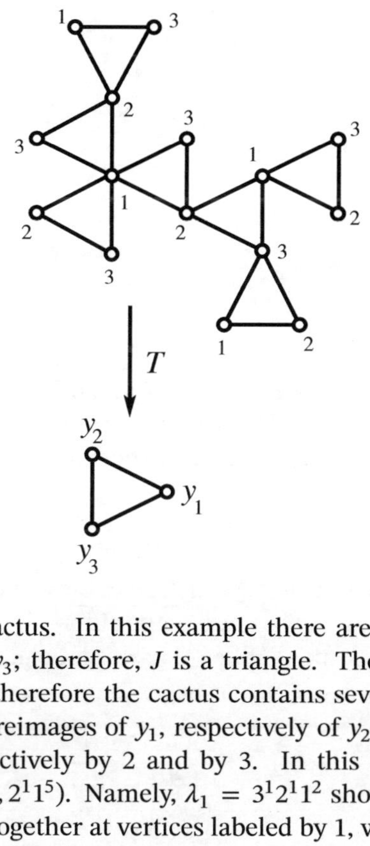

FIGURE 3.4. A cactus. In this example there are three finite critical values y_1, y_2, y_3; therefore, J is a triangle. The polynomial T is of degree 7, and therefore the cactus contains seven triangles. Vertices which are preimages of y_1, respectively of y_2 and of y_3, are labeled by 1, respectively by 2 and by 3. In this example we have $\Lambda = (3^1 2^1 1^2, 2^2 1^3, 2^1 1^5)$. Namely, $\lambda_1 = 3^1 2^1 1^2$ shows how many triangles are glued together at vertices labeled by 1, while partitions λ_2 and λ_3 correspond to labels 2 and 3.

The proof that condition (3.4) is also sufficient is divided into two parts. The first part is purely combinatorial and consists in constructing a cactus (at least one) with the vertex degrees corresponding to Λ. The second part is just a reference to Riemann's existence theorem which relates combinatorial data to the complex structure, as it was already the case in Proposition 2.12.

The proof of the *existence of a cactus* in question is similar to that of Proposition 3.4; namely, it consists in "cutting a leaf". Here a *leaf* means a copy of J which is attached to C at a single vertex (see Figure 3.5). This cutting operation must be carried out not with the cactus itself (since it is not yet constructed) but with its passport Λ: it is easy to verify that (3.4) implies that all partitions $\lambda_i \in \Lambda$ except maybe one contain a part equal to 1. We eliminate these parts, and diminish by 1 a part in the remaining partition. In this way we obtain a valid passport Λ' of degree $n-1$; then we construct inductively a smaller cactus; and then glue to it an nth copy of J. We leave details to the reader. □

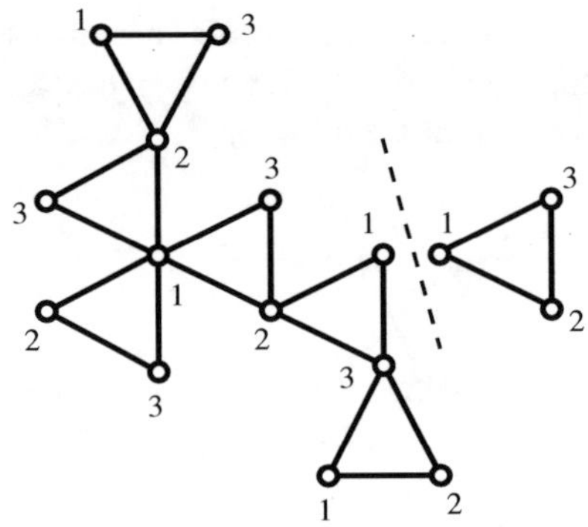

FIGURE 3.5. Cutting off a leaf from a cactus. A leaf exists since every partition in Λ, except maybe one, contains a part equal to 1: this is a consequence of (3.4).

Note that for rational functions, and even for Laurent polynomials, a similar statement is not valid, see [**Pak-09**]: conditions based on the Euler formula remain necessary but they are no longer sufficient. See also Theorem 3.1 (page 19) and Remark 3.12 (page 26).

Another approach to the proof of Proposition 3.8 is to use an enumerative formula due to Goulden and Jackson [**GoJa-92**] which gives the number of cacti corresponding to a given list of partitions Λ. Let us write a partition $\lambda \vdash n$ in the power notation: $\lambda = 1^{d_1}2^{d_2}\ldots n^{d_n}$ where d_i is the number of parts of λ equal to i, so that $\sum_{i=1}^n d_i = p$ (here p is the total number of parts in λ), and $\sum_{i=1}^n id_i = n$. Denote

$$N(\lambda) = \frac{(p-1)!}{d_1!\, d_2!\, \ldots\, d_n!}.$$

Then the following holds:

PROPOSITION 3.9 (Enumeration of cacti). *For a given* $\Lambda = (\lambda_1, \lambda_2, \ldots, \lambda_k)$ *satisfying conditions* (3.4) *we have*

$$\sum \frac{1}{|\mathrm{Aut}\,(C)|} = n^{k-2} \prod_{i=1}^{k} N(\lambda_i) \tag{3.5}$$

where the sum is taken over the cacti C with the passport Λ, and $|\mathrm{Aut}\,(C)|$ is the size of the automorphism group of C.

Now in order to prove Proposition 3.8 it suffices to remark that the right-hand side of formula (3.5) is always positive. □

REMARK 3.10 (Enumeration of ordinary trees). Taking $k = 2$ in Proposition 3.8 we may put the critical values y_1 and y_2 to 0 and 1, and replace the Jordan curve J passing through these points by the segment $[0, 1]$. Then a cactus becomes an *ordinary* bicolored plane tree with the passport $\Lambda = (\lambda_1, \lambda_2)$. The formula for this case was found earlier by Tutte [**Tut-64**]; it is a particular case of the above formula (3.5) for $k = 2$. The Tutte formula will be useful in the future: in order to verify that a given ordinary tree is a *unitree* we can just compute the number given by the formula and see if it is less than or equal to one (the number can be less than one when the tree in question admits non-trivial automorphisms).

3.6. Proof of the weak bound

Consider first the case $d = \gcd(\alpha, \beta) = 1$, so that $p + q > n + 1$. In this case inequality (3.3) is trivial: it is reduced to $\deg R \geq 0$. Thus, we only need to prove that this bound is attained.

We have $n + 1 \leq p + q \leq 2n$, or, equivalently, $-2n \leq -(p + q) \leq -n - 1$. Adding $2n + 1$ to both inequalities we get

$$1 \leq (2n + 1) - (p + q) \leq n.$$

Let us take an arbitrary partition $\lambda_3 \vdash n$ having $(2n + 1) - (p + q)$ parts, and also take $\lambda_1 = \alpha$ and $\lambda_2 = \beta$. Then for $\Lambda = (\lambda_1, \lambda_2, \lambda_3)$ conditions (3.4) are satisfied. Hence, there exists a polynomial $T(x)$ satisfying all the conditions of Proposition 3.8, with three critical values y_1, y_2, y_3 which may be chosen arbitrarily. Taking $P(x) = T(x) - y_1$ and $Q(x) = T(x) - y_2$ we obtain two polynomials which factorize as in (2.1) and whose difference is

$$R(x) = P(x) - Q(x) = y_2 - y_1 = \text{Const}.$$

Thus, the obvious lower bound $\deg R \geq 0$ is indeed attained.

Let us now consider the case $\gcd(\alpha, \beta) = d > 1$. In this case we must prove both the bound (3.3) and its attainability.

We have $P = f^d$ and $Q = g^d$. Therefore, $R = f^d - g^d$ may be factorized into d factors $f - \zeta g$, where ζ runs over the d-th roots of unity. If one of the factors, which we may without loss of generality assume to be $f - g$, has a degree $< n/d$, then the leading coefficients of f and g coincide. Hence, the leading coefficients of f and ζg for $\zeta \neq 1$ do not coincide, and all the remaining $d - 1$ factors have the degree exactly equal to n/d. This gives us the inequality

$$\deg R = \deg(f - g) + (d - 1) \cdot \frac{n}{d} \geq \frac{(d - 1)n}{d}.$$

According to the first part of this proof, the bound $\deg(f - g) \geq 0$ is attained, and therefore the bound (3.3) is also attained.

This finishes the proof of Theorem 3.7. □

Notice that the above reasoning may be used for deducing the attainabilty of the bound (1.6) in the case when condition (3.1) is satisfied and $d > 1$, from the case $d = 1$.

Indeed, it follows from the case of coprime α and β that

$$\min\deg(f-g) = \left(\frac{n}{d}+1\right)-(p+q),$$

and hence

$$\min\deg(f^d-g^d) = \left[\left(\frac{n}{d}+1\right)-(p+q)\right]+(d-1)\cdot\frac{n}{d} = (n+1)-(p+q).$$

EXAMPLE 3.11 (Weak bound). Let us take $n=6$, $\alpha=4^1 2^1$, and $\beta=2^3$, so that $p=2$, $q=3$, and $d=2$. Then we have

$$(n+1)-(p+q)=(6+1)-(2+3)=2$$

but this bound cannot be attained. The correct answer is given by Theorem 3.7:

$$\min\deg R = (d-1)\cdot\frac{n}{d} = (2-1)\cdot\frac{6}{2} = 3.$$

And, indeed, there is only one plane map with two black vertices of degrees 4 and 2, respectively, and with three white vertices of degree 2, see Figure 3.6. This map has two finite faces, but one of them is not of degree 1. The sum of degrees of the finite faces is 3.

FIGURE 3.6. This is the only plane map having the passport $(4^1 2^1, 2^3)$. We see that one of the faces is of degree 2.

REMARK 3.12 (Non-realizable planar data). Take α and β such that $\gcd(\alpha,\beta) = d>1$ and $\frac{n}{d}+1<p+q\le n+1$. Let us also take $r=(n+1)-(p+q)$ and $\gamma=(n-r,1^r)$. Then the passport $\pi=(\alpha,\beta,\gamma)$ satisfies the Euler relation: there are $p+q$ vertices, n edges, and $r+1$ faces, so that

$$(p+q)-n+(r+1) = (p+q)-n+[(n+1)-(p+q)+1] = 2.$$

However, a plane map with these data does not exist.

The principal vocation of this book is to use combinatorics for the study of polynomials. But here in this particular example we, essentially, deduce a non-trivial statement about plane maps from a trivial property of polynomials. Namely, we deduce the non-existence of certain maps from the fact that the degree of a polynomial cannot be negative.

CHAPTER 4

Recapitulation and perspectives

This short chapter is an intermediate stop on a long road. Its goal is to summarize what we have achieved so far and to draw the plans for the future, that is, for the subsequent chapters.

The general theory of dessins d'enfants implies that to every weighted bicolored plane tree there corresponds a Davenport–Zannier pair of polynomials. As a corollary, we obtained a lower bound for the degree of the difference of these polynomials, and showed that this bound is always attained (Theorems 3.3 and 3.7). This is a serious achievement, especially if we take into account that for number theorists it took 30 years to prove the same results. The merit goes to the pictorial representation of polynomials and rational functions.

Now, there are new adventures ahead of us: we shall study the Galois theory of the DZ-pairs. The coefficients of the DZ-polynomials are algebraic numbers. The universal (or the absolute) Galois group $\Gamma = \mathrm{Aut}(\overline{\mathbb{Q}}|\mathbb{Q})$ is the group of automorphisms of the field $\overline{\mathbb{Q}}$ of algebraic numbers. The group Γ acts on DZ-polynomials by replacing the coefficients by their algebraically conjugate ones. In this way, Γ also acts on weighted trees. The smallest field to which the coefficients belong is called the *moduli field* of the orbit. It is uniquely determined by the orbit and cannot be chosen arbitrarily.

A *combinatorial orbit* is a set of weighted trees having the same passport. By obvious combinatorial reasons, every combinatorial orbit is finite. Since the passport is a Galois invariant, every Galois orbit is a subset of a combinatorial orbit. The same statement may be reformulated as follows: every combinatorial orbit either is itself a Galois orbit, or it splits into several Galois orbits. See in this respect Example 9.23 where a combinatorial orbit of size 16 splits into four Galois orbits, of sizes 1, 2, 5 and 8, respectively.

The first case to study is the one of combinatorial orbits consisting of a single element. This innocuously looking problem is not simple at all, as the reader will see in the next chapter. The interest of this case is that the corresponding Galois orbit also consists of a single element, and therefore its field of moduli is "an extension of $\mathbb{Q}$ of degree 1". The latter means that this field is the field $\mathbb{Q}$ itself, so that the corresponding DZ-polynomials have rational coefficients. The polynomials themselves are computed in Chapter 6; the task is not simple either.

The next most important Galois invariant is the *monodromy group*; in certain publications, especially in Russian, this group is also called *edge rotation group*. This is a permutation group which acts on the edges of a dessin. It is generated by two permutations: the first one sends an edge to the next one, in the counter-clockwise direction, attached to the same black vertex; the second permutation does the same for the white vertex. Here, the most interesting result is the complete classification of primitive monodromy groups of weighted trees. Two chapters, 7 and 8, are devoted to this subject.

The monodromy group is a powerful and profound invariant but rather difficult to deal with. In Chapter 9 we consider some other Galois invariants. Almost all of them are related, in one or another way, to the monodromy group, but quite often they are easier to detect. Just think of a symmetry. A tree is symmetric if and only if its monodromy group has a non-trivial centralizer in the symmetric group. But in practice we do not need to compute either the group itself or its centralizer since the symmetry is visible with the naked eye. By the end of the chapter, however, we consider an invariant which is even more difficult to deal with than the monodromy group; it can be detected only "with a little bit of luck". It is related to the action of the *Hurwitz braid group* $\mathcal{H}_4$ on the quadruples of permutations. We call it *megamap invariant* since the Hurwitz space for the coverings with four ramification points is, in certain publications, called megamap. In other publications the same object is called *Fried family*.

In Chapter 10 we analyze a particular but very instructive example. The ultimate goal of the theory of dessins d'enfants was often formulated as follows: find a *complete set* of combinatorial Galois invariants (group-theoretic invariants were considered as a particular case of combinatorial ones). The example of Chapter 10 shows that this goal is unattainable since such a set of invariants simply does not exist. In this example, the combinatorial orbit consists of two elements and usually is defined over a quadratic field. The question of its splitting into two Galois orbits defined over $\mathbb{Q}$ is reduced to the well-known Pell equation. Other diophantine invariants were also known before but they were never considered as a specific class of invariants. Hence this approach to dessins d'enfants is still not well developed.

Chapter 11 is dedicated to the enumeration of weighted trees. It is not directly related to the principal contents of the book, but in practice it is often useful to know the size of a given combinatorial orbit.

The final (and very short) Chapter 12 presents a few directions of further research.

There is a long way to go in front of us, and we hope that the reader will get at least a part of a pleasure we have had while collecting this material.

CHAPTER 5

Classification of unitrees

The material of this chapter is based on the paper [**PaZv-14**] by Pakovich and Zvonkin. We give a complete classification of the passports and corresponding trees satisfying the following property: *a weighted bicolored plane tree having this passport is unique.* Such trees will be called *unitrees.* As we have explained before, in Proposition 2.20, Belyĭ functions corresponding to unitrees are defined over $\mathbb{Q}$. Therefore, the corresponding DZ-pairs consist of polynomials with rational coefficients.

Ordinary unitrees were classified by Adrianov as early as in 1989. However, this initial proof was never published since it looked too cumbersome. Later on, Adrianov [**Adr-07a**] proposed another proof based on Tutte's enumerative formula: see (3.5) for $k = 2$. The classification is found by carefully examining this formula and looking for the cases in which it gives a number ≤ 1 (recall that (3.5) counts each tree C with the weight $1/|\mathrm{Aut}(C)|$).

Our situation here is more difficult for two reasons. First, we deal with weighted trees; and, second, we don't have an enumerative formula at our disposal. To be more specific, we would need a formula which would give us, in an explicit way, the number of the weighted bicolored plane trees *corresponding to a given passport.* An additional difficulty here ensues from the fact that the same passport may correspond not only to trees but also to forests: see Chapter 11. Inclusion-exclusion approach is not well adapted to our goals.

Well... all these reasons must not discourage us to attack the problem.

ASSUMPTION 5.1 (Passports from now on). In the remaining part of the book we will consider only the passports (α, β) such that $\gcd(\alpha, \beta) = 1$ and $p + q \leq n + 1$ where p and q are the numbers of parts in α and β respectively.

According to Lemma 3.6, the case $\gcd(\alpha, \beta) > 1$ is reduced to this one. Indeed, starting from a tree $\mathcal{T}$ with $\gcd(\alpha, \beta) = d > 1$ we can obtain a tree $\widetilde{\mathcal{T}}$ with $\gcd(\widetilde{\alpha}, \widetilde{\beta}) = 1$ by dividing the weights of all edges of $\mathcal{T}$ by d, and it is easy to see that $\mathcal{T}$ is a unitree if and only if $\widetilde{\mathcal{T}}$ is a unitree.

Recall that the face partition γ is determined by (α, β) and is always equal to $\gamma = (n - r, 1^r)$ where $r = (n + 1) - (p + q)$.

DEFINITION 5.2 (Unitree). A weighted bicolored plane tree such that there is no other weighted bicolored plane trees with the same passport, is called a *unitree.*

5.1. Statement of the main result

DEFINITION 5.3 (Diameter). The *diameter* of a tree is the length of the longest path in this tree.

The classification of unitrees is summarized in the following theorem.

THEOREM 5.4 (Complete list of unitrees). *Up to an exchange of black and white colors and to a multiplication of all the weights by* $d > 1$*, the complete list of unitrees consists of the following* 20 *cases:*

- *Five infinite series A, B, C, D, E of trees shown in* Figures 5.1, 5.2, 5.3, 5.4, and 5.5, *involving a number of degree parameters and two weight parameters* s *and* t *which are supposed to be coprime (therefore, either* $s \neq t$*, or* $s = t = 1$*). Note that*
  - *for the diameter* ≥ 5*, only the trees of the types B and E exist;*
  - *for the diameter* 4*, the trees of types B, D, and E exist;*
  - *for the diameter* 3*, the trees of types B, C, and E exist.*
- *Five infinite series F, G, H, I, J without weight parameters but with degree parameters; they are shown in* Figures 5.6 *and* 5.7.
- *Ten sporadic trees K, L, M, N, O, P, Q, R, S, T shown in* Figures 5.8, 5.9, 5.10 *and* 5.11.

REMARK 5.5 (Non-disjoint). The above series are not disjoint. For example, the trees of the series C with $k = l = 1$ also belong to the series B. If $s > t$ then the case C becomes a part of E_3, up to a renaming of variables; etc.

REMARK 5.6 (Ordinary unitrees). The list of ordinary unitrees consists of the following cases: the series A, B and C with $s = t = 1$; the series F, H and I; and the sporadic tree Q.

REMARK 5.7 (White vertices of degree 2). Notice that quite a few of our trees have all their white vertices being of degree 2, and thus, according to Convention 2.15, we can make these vertices implicit and draw the pictures as usual (i. e., non-bicolored) plane maps. The corresponding maps are shown in Figure 5.12.

The strategy of the proof of Theorem 5.4 is as follows. We propose various transformations of trees changing the trees themselves while preserving their passports: this is a way to show that the combinatorial orbit of a given tree consists of more that one element. The trees which survive such a surgery are (a) those to which the transformation in question cannot be applied, and (b) those for which the transformed tree turns out to be isomorphic to the initial one. In this way we gradually eliminate all the trees which are not unitrees. Then, at the final stage, we show that all the trees which have passed through all the sieves are indeed unitrees.

The proof ends on page 56. It is rather long and, we must admit, may give you a pain in the neck. We would be glad to have a more elegant proof but for the time being there is none.

DEFINITION 5.8 (Rooted tree). A tree with a distinguished leaf edge is called *rooted tree*, and the distinguished edge itself, independently of its weight, is called the *root* of the tree. Two rooted trees are isomorphic if there exists an isomorphism which sends the root of one of the trees to the root of the other one.

DEFINITION 5.9 (Branch). Let a vertex v of a tree $\mathcal{T}$ be given. Then a *branch* of $\mathcal{T}$ attached to v is a rooted tree which is a subtree of $\mathcal{T}$ containing a single edge incident to v, and this edge serves as the root of the subtree.

The following characterization of unitrees eliminates a vast amount of possibilities.

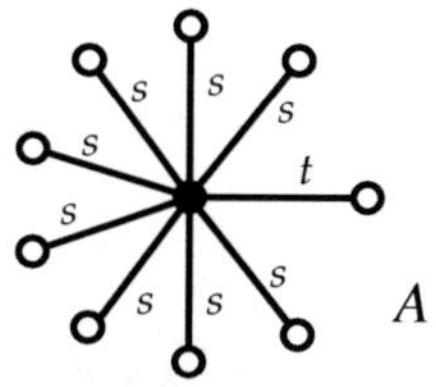

FIGURE 5.1. Series A: stars. There are $k \geq 0$ edges of the weight s and only one edge of the weight t which may be equal or not to s (recall that if they are equal then $s = t = 1$).

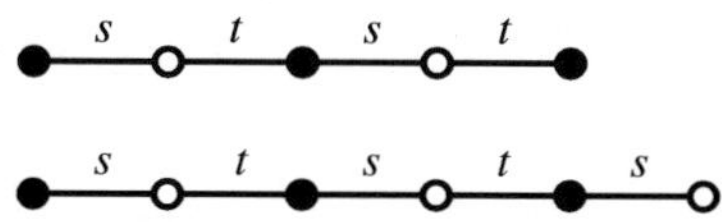

FIGURE 5.2. Series B: periodic chains of arbitrary length. We distinguish the chains of even and odd length since they have passports of different types.

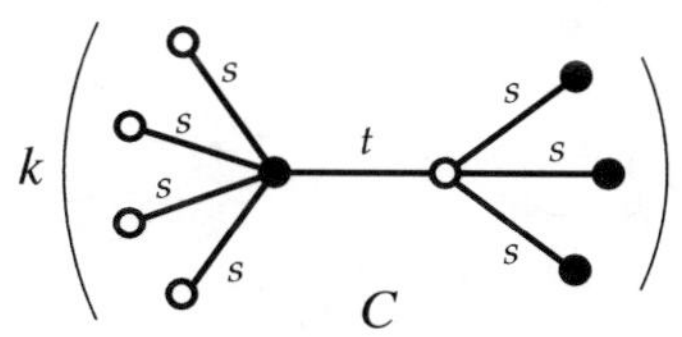

FIGURE 5.3. Series C: brushes of diameter 3. Here $k, l \geq 1$.

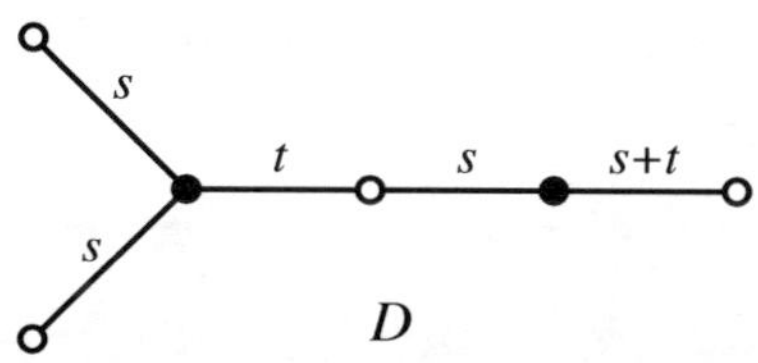

FIGURE 5.4. Series D: brushes of diameter 4. There are exactly two leaves of weight s and exactly one leaf of weight $s + t$.

LEMMA 5.10 (Branches of a unitree). *All branches going out of a vertex of a unitree, except maybe one, are isomorphic as rooted trees. This property must be true for every vertex of a unitree.*

PROOF. Let us call a vertex *central* if it is obtained by the following procedure. We cut off all the leaves of the tree; then do the same with the remaining smaller tree, then again, etc. In the end, what remains is either a single vertex, or an edge. In the first case, there is a single central vertex; in the second case, there are two of them, one black and one white. Obviously, *any isomorphism of trees sends a central vertex to a central one*, and if there are two central vertices, it sends the black central vertex to the black one, and the white, to the white one. On the other hand, according to Definition 2.23,

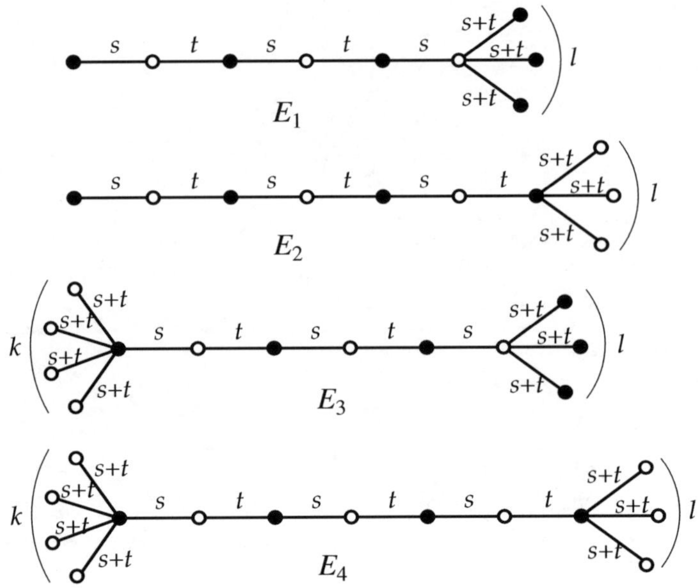

FIGURE 5.5. Series E: brushes of an arbitrary length. If there is a leaf of weight s, as in two upper pictures, then this leaf it is "solitary" on one of the ends of the brush; otherwise, all the leaves are of the weight $s+t$. The parameters $k, l \geq 1$.

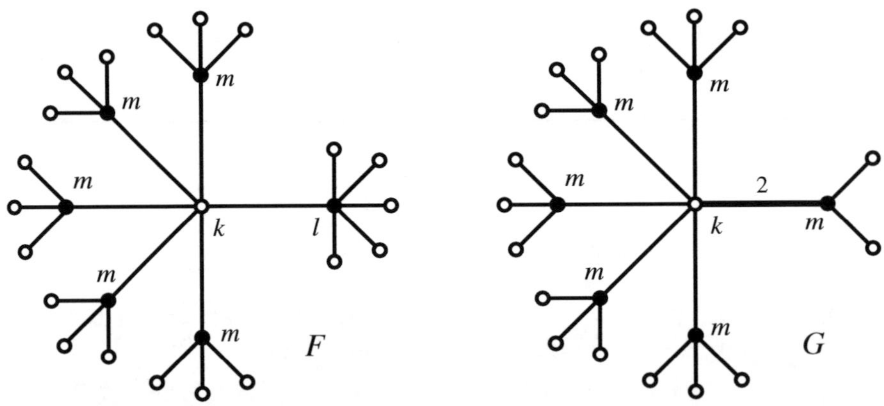

FIGURE 5.6. Two series of unitrees of diameter 4. In the trees of the series F all the edges are of weight 1; the degrees of vertices (except the leaves) are indicated. In the trees of the series G, there is exactly one edge of weight 2, all the other edges being of weight 1; note that for the series G the degrees of the black vertices are all equal.

any isomorphism which sends a vertex v to itself must preserve the cyclic order of the branches attached to v. Thus, the property affirmed in this lemma, namely, that all the branches except maybe one are isomorphic, is valid for a central vertex or vertices,

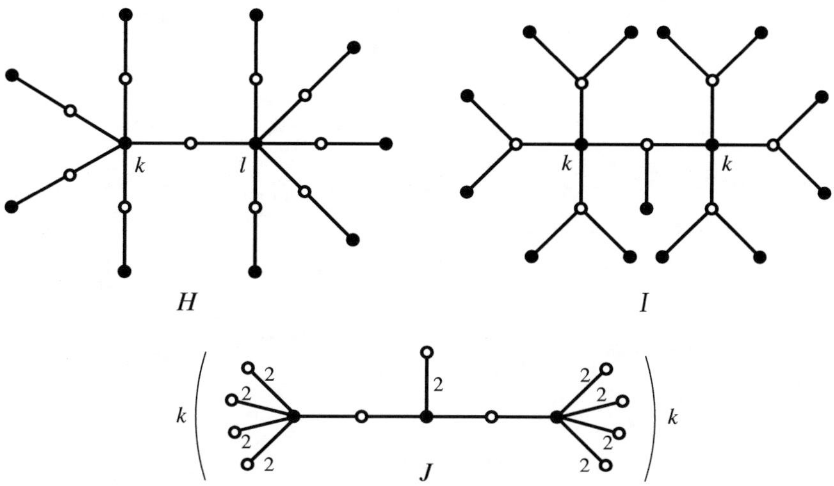

FIGURE 5.7. Three series of unitrees of diameter 6. In H and I, all edges are of weight 1. In H the black vertices which are not leaves are of degrees k and l which may be non-equal; in I they are of the same degree k. In H all white vertices are of degree 2; in I, they are all of degree 3. In J, the number of leaves of the weight 2 on the left and on the right is the same.

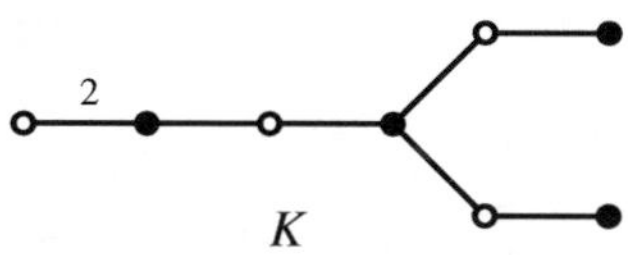

FIGURE 5.8. A sporadic unitree of diameter 5.

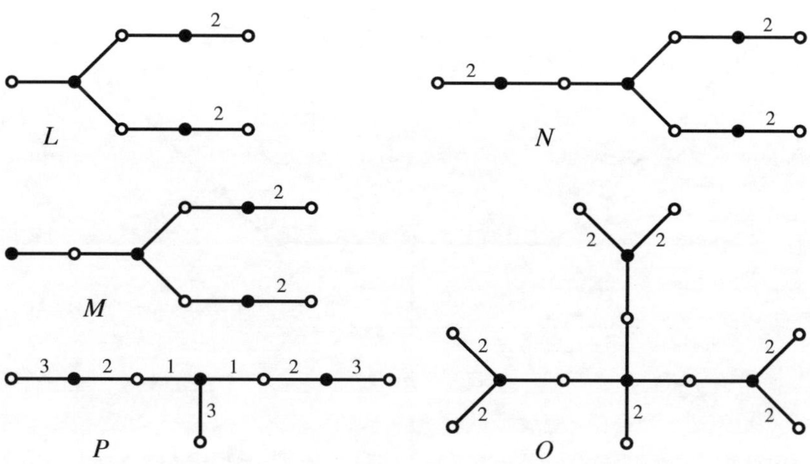

FIGURE 5.9. Five sporadic unitrees of diameter 6. Notice that the tree N is symmetric, with the symmetry of order 3. (We have drawn it in a non-symmetric way in order to save space.)

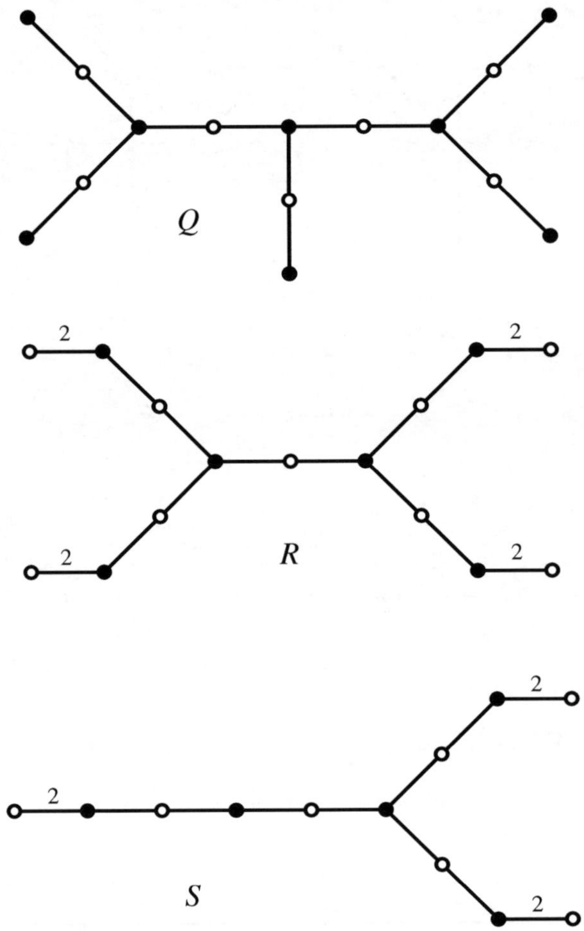

FIGURE 5.10. Three sporadic unitrees of diameter 8.

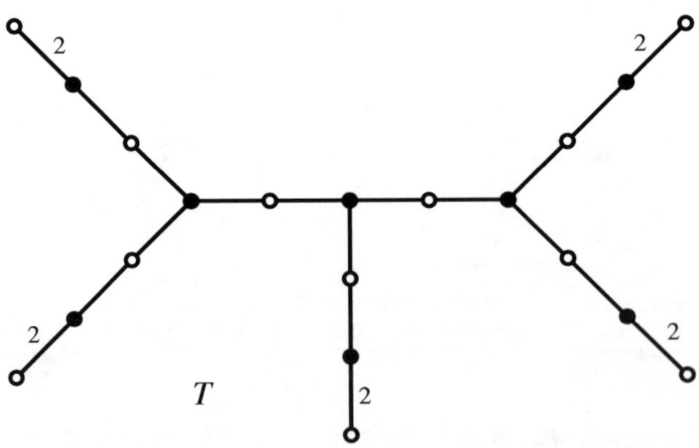

FIGURE 5.11. A sporadic unitree of diameter 10.

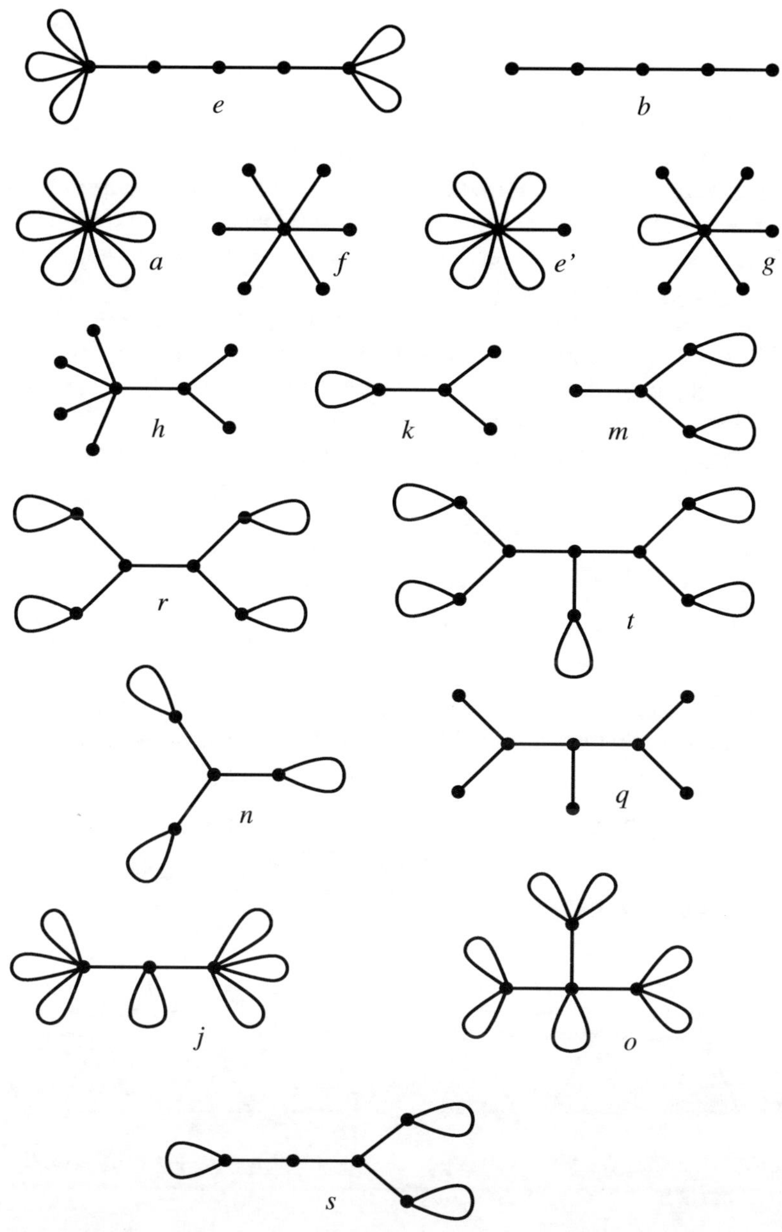

FIGURE 5.12. Unimaps. Small letters correspond to the capital letters by which we have previously denoted the unitrees; for example, the series e here is a particular case of the series E (when $s = t = 1$). Note that a, b, e', f, and g are particular cases of e. Note also that the series h and j and the sporadic unimaps k, m, n, o, q, r, s are not "particular cases" but just coincide with H, J, K, etc., respectively. (Important remark: *we do not pretend that there are no other unimaps. The above ones are those which correspond to unitrees.*)

since otherwise an operation of *exchanging* two non-isomorphic branches attached to the central vertex would change the cyclic order of branches around this vertex.

Further, observe that the operation of exchanging two non-isomorphic branches attached to a vertex of a *rooted* tree always changes this tree unless all the branches attached to this veretx, except maybe the branch containing the root, are isomorphic. Indeed, the introduction of a root makes a cyclic order on branches around any vertex into a *linear order* on the branches incident to it and not containing the root. The only possibility to make this linear order invariant under the operation of exchanging the branches is to make them all equal.

Now, if a vertex v of a tree $\mathcal{T}$ is not central then it belongs to a branch $\mathcal{V}$ attached to a central vertex. This branch itself is a rooted tree and, in case if the condition of the lemma is not satisfied, the operation of exchanging of the branches changes $\mathcal{V}$. However, changing the branch $\mathcal{V}$ would mean changing a single branch of $\mathcal{T}$ attached to its central vertex. This would make $\mathcal{T}$ not isomorphic to itself. Thus, in this case $\mathcal{T}$ would not be a unitree. □

5.2. Weight distribution

Sometimes, we can change not the topology of the tree but the distribution of its weights, while remaining in the same combinatorial orbit. Let us first formulate a statement which is entirely obvious:

LEMMA 5.11 (Weight distribution). *If a passport (α, β) corresponds to a unitree then the corresponding weight distribution μ* (see Definition 2.22, page 17) *is determined by α and β in a unique way.* □

Now, another observation:

LEMMA 5.12 (Condition on weights). *Let s, t, u be the weights of three successive edges of a unitree, as in* Figure 5.13, left. *If $s \leq u$ then either $u = s$, or $u = s + t$. Similarly, if $s \geq u$ then either $u = s$, or $u = s - t$.*

FIGURE 5.13. Weight exchange. If $s < u$ but $u \neq s + t$ then exactly two parts of the weight distribution μ have changed.

PROOF. Rotating if necessary the tree under consideration, without loss of generality we may assume that $s \leq u$. If $s < u$ then we can construct the tree shown in Figure 5.13, right, replacing the weight u with $s + t$, the weight t with $u - s$, and exchanging the places of the subtrees $\mathcal{B}$ and $\mathcal{D}$. We see that the vertex degrees of the new tree are the same as in the initial one, while the weights of two edges have changed, unless $u = s + t$. Thus, if $s < u$ but $u \neq s + t$, then exactly two parts of the weight distribution μ have changed, which contradicts Lemma 5.11. □

COROLLARY 5.13 (Adjacent edges). *If in a unitree there are two adjacent edges of the same weight s, and at least one of them is not a leaf, then $s = 1$.*

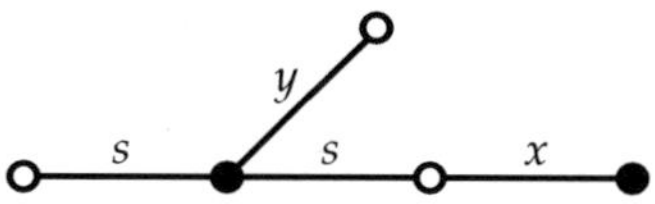

FIGURE 5.14. Two adjacent edges of the same weight s; one of them is not a leaf.

PROOF. An edge which is not a leaf must be the middle edge of a path of length 3, see Figure 5.14.

According to Lemma 5.12 we have either $x = s$ or $x = 2s$. If there is an edge of weight y attached to the middle edge of the path, like in Figure 5.14, then we have either $y = x$ or $y = x + s$, so the possible values for y are s, $2s$, or $3s$. Dealing with the other edges of the tree in the same way we see that the weights of all of them are multiples of s. According to Assumption 5.1 this means that $s = 1$. □

PROPOSITION 5.14 (Path s, t, s). *Suppose that a unitree contains a path of three successive edges having the weights s, t, s. Then the only possible weights for all the edges of this tree are s, t, or $s + t$.*

PROOF. Let us make an operation inverse to the one used in the proof of Theorem 3.5 (page 20), that is, "rip" the tree along the edge of the weight t, as in Figure 5.15.

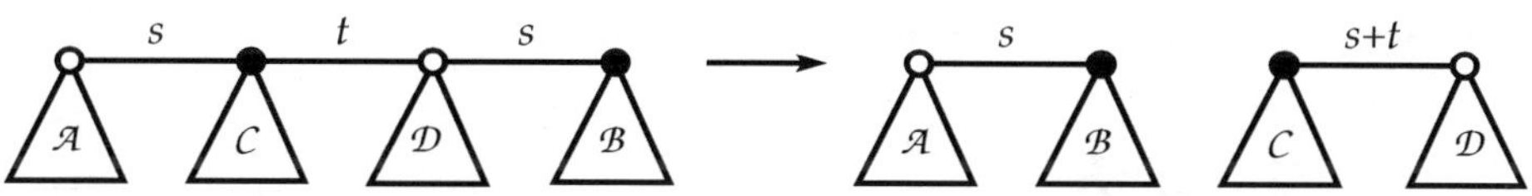

FIGURE 5.15. "Ripping" a tree: an operation inverse to that of the proof of Theorem 3.5.

We now suppose that in one of the subtrees $\mathcal{A}, \mathcal{B}, \mathcal{C}, \mathcal{D}$ there exists an edge of a weight $x \neq s, t, s + t$, and will try to stitch the two trees together in a different way.

(1) Suppose that in one of the subtrees $\mathcal{A}$ or $\mathcal{B}$ there exists an edge of a weight $x \neq s + t$. Stitch it to the edge of the weight $s + t$ by the procedure explained in the proof of Theorem 3.5.
   (a) If $x < s + t$ then the weights of the four edges participating in the operations of ripping and stitching, instead of being s, t, s, x become $s + t - x$, x, x, s. Removing from the two sets the coinciding elements s and x, we get, on the one hand, s, t, and, on the other hand, $s + t - x, x$. These sets coincide only when $x = s$ or $x = t$.
   (b) If $x > s + t$ then, instead of s, t, s, x, we obtain $x - s - t, s + t, s + t, s$. These two sets cannot coincide at all since x is greater than every term in the second set.

(2) Suppose that in one of the subtrees $\mathcal{C}$ or $\mathcal{D}$ there exists an edge of a weight $x \neq s$. Stitch it to the edge of the weight s by the same procedure as above.
   (a) If $x > s$ then, instead of s, t, s, x, we get $x - s, s, s, s + t$. Removing s, s from both sets we get, on the one hand, t, x, and on the other hand, $x - s, s + t$. These sets coincide only when $x = s + t$.

(b) If $x < s$ then the new set of weights is $s - x, x, x, s + t$; this set cannot coincide with s, t, s, x since $s + t$ is greater than every term in the second set.

Thus, the hypothesis that there exists an edge of a weight $x \neq s, t, s + t$ leads to a contradiction. The proposition is proved. □

REMARK 5.15. Notice that the operation of ripping and stitching introduced in the proof of Proposition 5.14 often leads to another tree even in the case when the weights of all the edges of the tree under consideration are s, t, or $s + t$. Below we will often use this operation and call it *sts*-operation.

PROPOSITION 5.16 (Path $s, t, s+t$, I). *Suppose that a unitree contains a path of three successive edges having the weights s, t, $s + t$, and suppose also that $s \neq t$. Then the edge of the weight $s + t$ is a leaf.*

PROOF. Take the tree shown in Figure 5.16, left, and exchange the subtrees $\mathcal{B}$ and $\mathcal{D}$. Obviously, both trees in the figure have the same passport. Suppose that the edge of the weight $s + t$ is not a leaf, so that the subtree $\mathcal{D}$ of the left tree is non-empty. According to Lemma 5.10, the edges of $\mathcal{D}$ which are adjacent to the vertex q have the same weight. Denote this weight by x. By Lemma 5.12, the possible values of x are either t or $s + 2t$. In the first case, we get a sub-path containing three edges of the weights $t, s + t, t$, and, according to Proposition 5.14, the only possible edge weights in such a tree could be t, $s + t$, and $s + 2t$. But this contradicts the supposition that we have already an edge of the weight s with $s \neq t$.

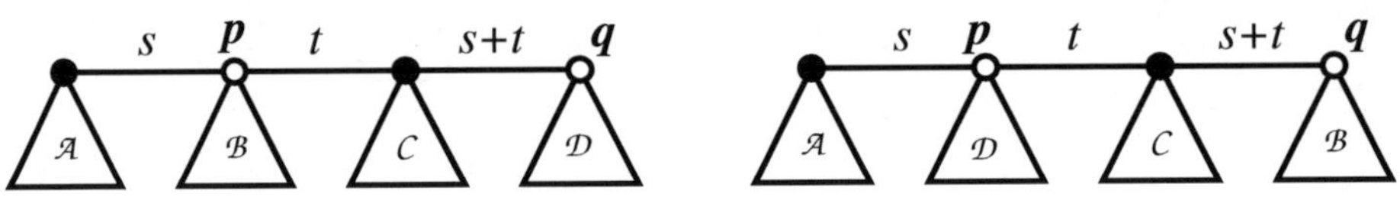

FIGURE 5.16. Suppose that the subtree $\mathcal{D}$ is not empty.

If, on the other hand, $x = s + 2t$, then there are at least three *non-isomorphic* subtrees attached to the vertex p in the right tree, since there are edges of three different weight s, t, and $s + 2t$ incident to this vertex. This situation violates Lemma 5.10. □

REMARK 5.17 ($s = t$). If $s = t$ then both arguments in the above proof are no longer valid, though it is still difficult to make left and right trees of Figure 5.16 isomorphic. (Recall that if $s = t$ then $s = t = 1$, see Corollary 5.13.) However, it is possible to construct unitrees having the paths of length 4 with the weights $1, 1, 2, 1$ (see the series G, Figure 5.6, page 32), and the paths with the weights $1, 1, 2, 3$ (see the tree P, Figure 5.9, page 33).

PROPOSITION 5.18 (Path $s, t, s + t$, II). *Suppose that a unitree contains a path of three successive edges having the weights s, t, $s+t$, and suppose also that the vertex incident to the edges of weights s and t has the valency $s + t$. Then the edge of the weight $s + t$ is a leaf.*

PROOF. Keeping notation of Figure 5.16, it is enough to observe that if the subtree $\mathcal{B}$ is empty, then the two trees cannot be isomorphic since the right one has more leafs than the left one. This statement remains valid also for $s = t$. □

5.3. Brushes

A brush is a chain with two bunches of leaves attached to its ends: see formal definition below. Typical representatives of brushes are the trees shown in Figure 5.5, page 32.

In this section we classify all brush unitrees.

DEFINITION 5.19 (Crossroad). A vertex of a tree is *profound* if, after all the leaves being removed from the tree, this vertex does not become a leaf. A vertex of a tree is a *crossroad* if it is profound and has at least three branches going out of it.

DEFINITION 5.20 (Brush). A tree is called a *brush* if it does not contain crossroads.

PROPOSITION 5.21 (Brush unitrees). *A brush unitree belongs either to one of the series A, B, C, D, E* (Figures 5.1, 5.2, 5.3, 5.4, and 5.5, pages 31–32), *or to the series F with the degree of the central vertex $k = 2$, or to the series G with the degree of the central vertex $k = 3$* (Figure 5.6, page 32).

PROOF. In this section we present only a part of the proof, namely, we eliminate brush trees which are *not* unitrees. The uniqueness of the remaining brush trees will be proved later, in Section 5.7.

When all the edges of a tree are leaves we get the series A consisting of stars (Figure 5.1). According to Lemma 5.10, only one of the leaves may have a weight different from the others.

For the trees of diameter three, Lemma 5.12 leads to two possible patterns. One of them corresponds to the series C (Figure 5.3); the other one is shown in Figure 5.17, left. We see that when $k > 1$ we can transform the left tree of Figure 5.17 into the right one. The new tree has the same passport but is not isomorphic to the initial one. Thus, the pattern shown in Figure 5.17, left, is not a unitree. If $k = 1$ then this pattern is a particular case of the series E_1, so it is a unitree.

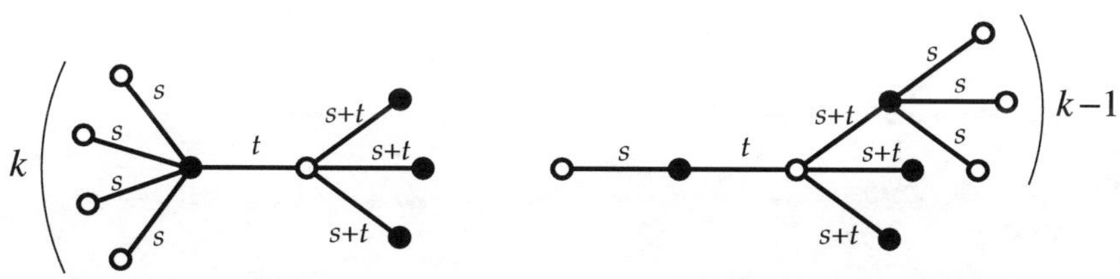

FIGURE 5.17. If $k > 1$, the tree on the left can be transformed into the one on the right. The new tree is different from the initial one since it has a diameter 4 instead of 3.

Now let us consider first the trees of diameter ≥ 5, and after that return to the diameter 4 case. Suppose that a tree contains two adjacent edges of weights s and t which are not leaves. It follows from Lemma 5.12 and Proposition 5.18 that the weights s and t alternate along all the path connecting vertices from which the leaves grow. Furthermore, since this path contains at least three edges, it follows from Proposition 5.14 that the only possible weight of a leaf which is not obtained by the further alternance of s and t is $s+t$. Now look at Figure 5.18, where an *sts*-operation is applied to a brush tree having $k \geq 2$ leaves of weight s on one of its ends. The tree thus obtained, shown on the right, is distinct from the initial one since it contains a crossroad. A similar surgery can be made if there are $l \geq 2$ leaves of the weight s or t (depending on the parity of the diameter) on the right end of the tree. Thus, for the diameters ≥ 5 only the types B and E survive. Namely, if a bunch of leaves at an end of the tree contains two or more leaves then the weight of these leaves is $s + t$.

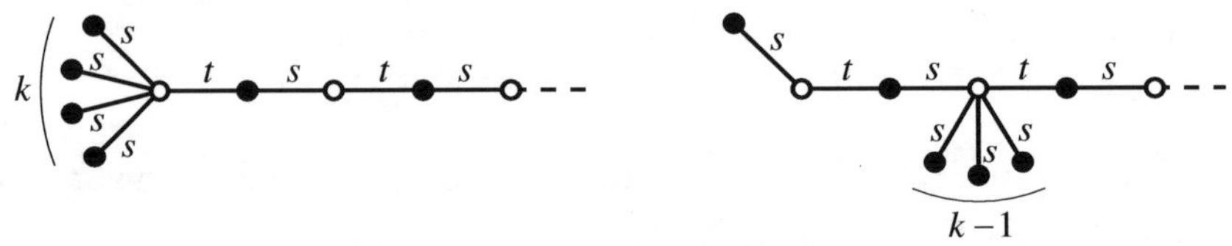

FIGURE 5.18. These trees have the same passport. They are not isomorphic since the right tree contains a crossroad while the left one does not. Therefore, if $k \geq 2$ then the weight of the leaves must be $s + t$ and not s or t.

The above argument fails for the brush trees of diameter 4: indeed, this time the operation shown in Figure 5.18 does not create a crossroad. Therefore, the diameter 4 case needs a special consideration. Let us take a diameter of the tree, that is, a chain of length 4 going from one of its ends to the other. By Lemma 5.12 the sequence of the weights of its three first edges can be either s, t, s, or s, t, $s + t$, or $s + t$, t, s. Consider first the case s, t, $s + t$, so that the edge of the weight $s + t$ is not a leaf. In this case, by Proposition 5.16 we necessarily have $s = t = 1$ and a tree either belongs to the series G, where the degree of the central vertex is $k = 3$, or has the form shown in Figure 5.19 on the left. In the last case, however, the tree under consideration is not a unitree, which can be seen by a transformation shown on the right.

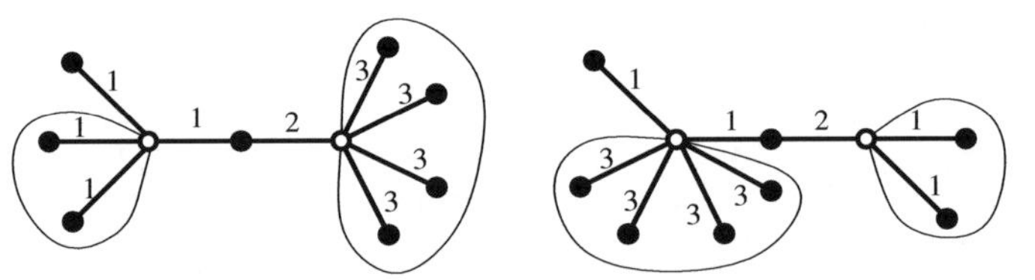

FIGURE 5.19. The bunch of leaves of weight 1 transplanted from left to right can be empty, if there is only one leaf on the left.

Assume now that the sequence of weights starts with s, t, s. Using Lemma 5.12 again we see that it must be a part of one of the following three possible sequences: either s, t, s, t, or $s, t, s, s+t$, or $s, t, s, t-s$. For the latter one, taking $s' = t - s$, $t' = s$, we find the already considered above case $s', t', s' + t'$ if read from right to left. Two other forms are shown in Figure 5.20 on the left.

The first of these forms (above, left) can be transformed in the way shown on the right. The new tree is not equal to the initial one, unless either it belongs to the type B (that is, $k = l = 1$), or $s = t = 1$, which is a particular case of the series F, where the degree of the central vertex is $k = 2$. For the second form (below, left), the operation

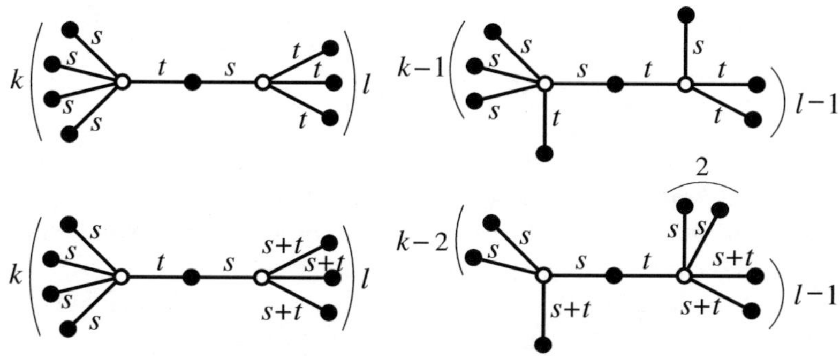

FIGURE 5.20. The trees on the upper level have the same passport; they are different if $s \neq t$ and at least one of the parameters k, l is greater than 1. The trees on the lower level also have the same passport; they are different if either $k > 2$, or $l > 1$, or both.

shown on the right cannot be applied when $k = 1$, that is, when the tree belongs to the series E_1; and it does not change the tree when $k = 2$ and $l = 1$, which corresponds to the series D.

Finally, if the sequence of weights of edges of a diameter starts as $s + t$, s, t, then either a tree belongs to the series E_4 (or E_2), or the sequence of weights of the edges of a diameter is $s + t$, s, t, $s - t$. In the latter case, however, Proposition 5.16 yields that $s - t = t = 1$, implying that a tree is the one shown in Figure 5.19 on the left. □

5.4. Trees with repeating branches of height 2

From now on we will assume that the trees we consider are not brushes, that is, they contain at least one crossroad. Recall that a crossroad is a profound vertex at which three or more branches meet, see Definition 5.19. A typical tree with a crossroad is shown in Figure 5.21.

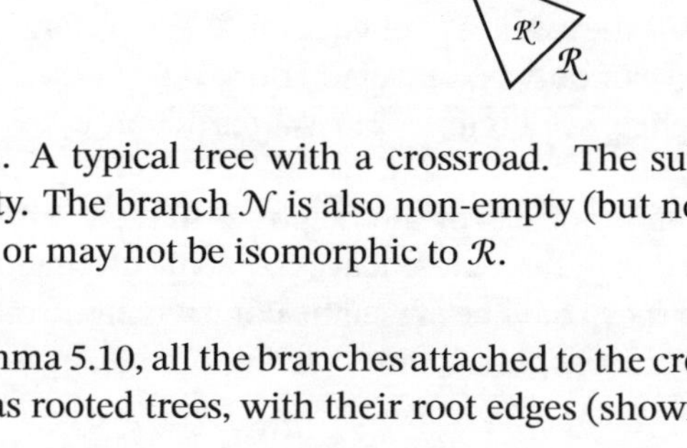

FIGURE 5.21. A typical tree with a crossroad. The subtrees $\mathcal{R}'$ are all non-empty. The branch $\mathcal{N}$ is also non-empty (but not necessarily $\mathcal{N}'$). $\mathcal{N}$ may or may not be isomorphic to $\mathcal{R}$.

According to Lemma 5.10, all the branches attached to the crossroad, except maybe one, are isomorphic as rooted trees, with their root edges (shown by thick lines in the figure) being incident to the crossroad. We call these branches *repeating*; in the figure they are denoted by the same letter $\mathcal{R}$; the subtrees of $\mathcal{R}$ attached to the root are denoted by $\mathcal{R}'$. The subtrees $\mathcal{R}'$ are non-empty since otherwise the vertex to which $\mathcal{R}$ and $\mathcal{N}$ are attached would not be profound. The number of branches of the type $\mathcal{R}$ can be two or more, but the majority of the transformations given below involve only two branches; therefore, in the majority of pictures we will draw only two repeating branches.

By convention, we suppose that *the branch $\mathcal{N}$ is always non-empty*. If all the branches meeting at the crossroad are isomorphic to $\mathcal{R}$ and might therefore be all considered as repeating, we take an arbitrary one and, somewhat artificially, declare it to be the "non-repeating" branch $\mathcal{N}$. The subtree $\mathcal{N}'$ has a right to be empty.

The roots of repeating branches are adjacent edges which are not leaves. Therefore, according to Corollary 5.13, their weights must be equal to 1. Finally, the *height* of a repeating or non-repeating branch is the distance from its root vertex (that is, the crossroad) to its farthest leaf.

In the previous section we classified the brush unitrees, which are by definition unitrees without crossroads. In this section we establish a complete list of all possible unitrees whose repeating branches all have the height 2. More precisely, we assume that for *any* crossroad of a unitree under consideration the repeating branches are of height 2.

PROPOSITION 5.22 (Repeating branches of height 2). *A unitree whose all repeating branches are of height 2 belongs to one of the types F, G, H, or K.*

PROOF. The proof is long: it ends on page 45. First of all observe that weights of leaves of repeating branches cannot be equal to 2 since otherwise the transformation shown in Figure 5.22 would lead to a non-isomorphic tree (the tree on the right has fewer leaves than the one on the left).

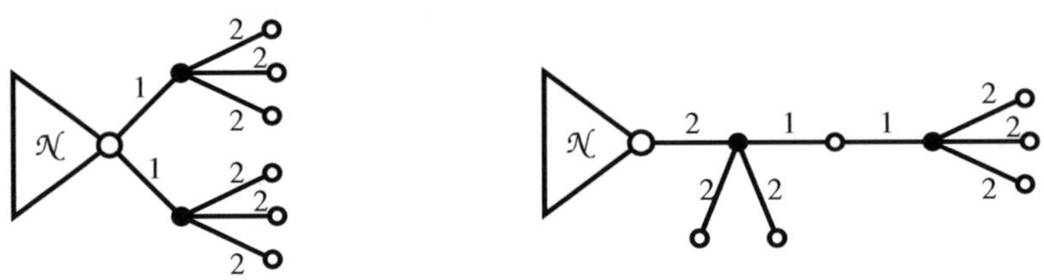

FIGURE 5.22. A transformation of repeating branches of height 2 with leaves of weight 2.

Thus, the weights of these leaves are equal to 1. Therefore, according to Lemma 5.12 the root edge of $\mathcal{N}$ can only be of weight 1 or 2. The case $\mathcal{N}' = \emptyset$ (that is, when $\mathcal{N}$ consists of a single edge, see Figure 5.21) is a particular case of the series F and G. Suppose then that $\mathcal{N}' \neq \emptyset$.

If the non-repeating branch is of the height 2, that is, if it is a root edge with a bunch of leaves attached to it, then these leaves could be of weight 1 or 2 when the root edge is of weight 1, and they could be of weight 1 or 3 when the root edge is of weight 2. However, the leaves of the weights 2 and 3 are impossible, as two transformations of Figure 5.23 show.

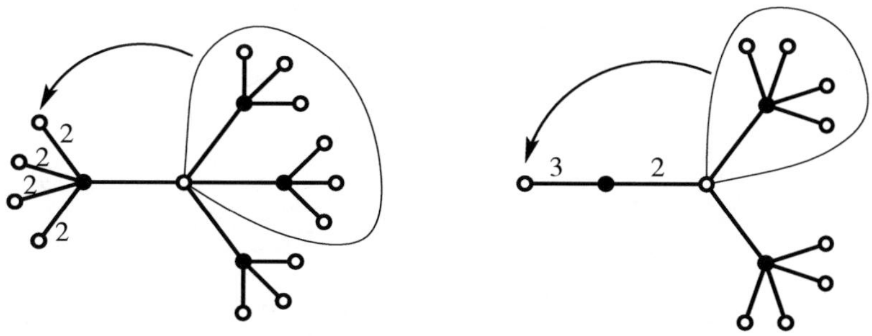

FIGURE 5.23. Non-repeating branch of height 2. These transformations show that the weight of its leaves cannot be 2 or 3.

In this figure, we take all the repeating branches except one and re-attach them to one of the leaves of the non-repeating branch. These transformations always change the trees: on the left, there appears a non-leaf of weight 2, and on the right, there appears a non-leaf of weight 3. Thus, all the leaves of the non-repeating branch must be of weight 1.

In addition, when the root edge of the non-repeating branch is of weight 2 this branch cannot be isomorphic to the repeating branches. This observation implies that the degree of the black vertex lying on this branch must be equal to the degrees of the black vertices of repeating branches since otherwise an exchange of leaves between repeating and non-repeating branches could be possible. Thus, the only remaining possibilities are the trees of the types F and G, see Figure 5.6 (page 32).

Consider now the case of a non-repeating branch of height ≥ 3. First suppose that the vertex q, which is the nearest neighbor of the crossroad vertex p when we move along the non-repeating branch, is itself a crossroad. According to our supposition, the tree does not have repeating branches of the height greater than 2. Hence, the repeating branches growing out of q are of height 2. But then a leaf $\mathcal{L}$ of such a branch can be interchanged with a repeating branch $\mathcal{U}$ growing out of p, see Figure 5.24. This operation would create at least three different trees attached to p: one of them would be of height 1, another one, of height 2, and the third one, of height 4. Thus, this possibility is ruled out.

Suppose next that the vertex q is not a crossroad. Then the tree looks like the one in Figure 5.25, top left, where $\mathcal{A}$ is non-empty. *A priori*, there are four possibilities for the values (s, t), namely, $(1, 1)$, $(1, 2)$, $(2, 1)$, and $(2, 3)$. The case $(2, 3)$ can be immediately ruled out since the edge of weight 3 should be a leaf by Proposition 5.16, but we have supposed that $\mathcal{A} \neq \emptyset$.

The cases $(s, t) = (1, 1)$ or $(2, 1)$ can be treated together. When $t = 1$ we can re-attach $\mathcal{A}$ to one of the leaves of the repeating branches, as is shown in the same figure on the top right. Among the branches attached to the vertex p of the tree thus obtained there is only one branch of a height greater than 2: it is $\mathcal{W}$. Therefore, all the remaining branches are repeating, so we may conclude that $s = 1$ (the case $s = 2$ is impossible), and all the repeating branches have only one leaf. Thus, the tree looks like the one on the bottom left in Figure 5.25, where two possibilities may occur: either $u = 1$; or

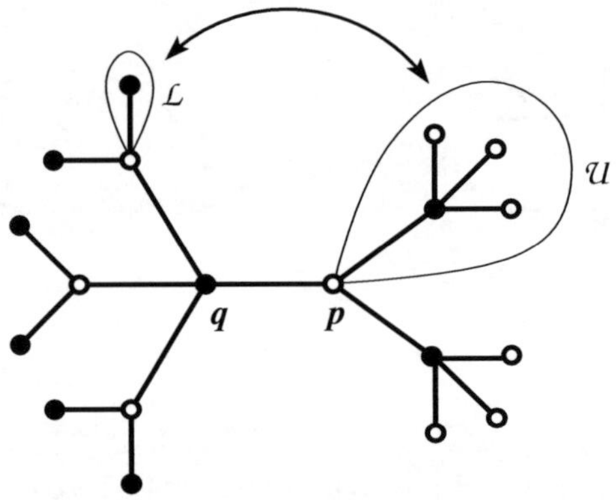

FIGURE 5.24. A leaf $\mathcal{L}$ can be interchanged with a repeating branch $\mathcal{U}$.

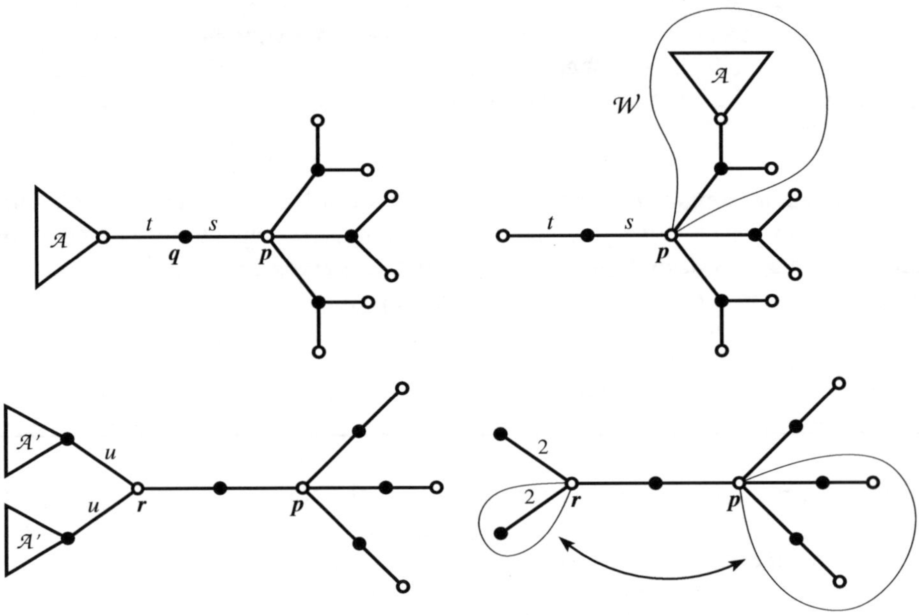

FIGURE 5.25. Illustration to the proof of Proposition 5.22.

$u = 2$, and then, according to Proposition 5.18, $\mathcal{A}' = \emptyset$, so the edges of the weight $u = 2$ are leaves.

In the first case, we can exchange the repeating branches attached to the vertices p and r. Therefore, they all must be equal, and we get a tree of the type H, see Figure 5.7 (page 33). In the second case, we can interchange one of the leaves of weight 2 with two repeating branches attached to p. The only tree which does not change after this transformation is the one which has exactly one leaf of weight 2 and exactly two repeating branches attached to p, that is, the tree K, see Figure 5.8 (page 33).

There remains the last case to be ruled out: when the tree shown in Figure 5.25, top left, has $s = 1$ and $t = 2$; see also Figure 5.26, left. In this case we can interchange the subtree $\mathcal{A}$ with all but one repeating branches (see the tree on the right of Figure 5.26). We see that $\mathcal{A}$ must consist of several copies of the branch $\mathcal{U}$ since otherwise $\mathcal{A}$ should consist of copies of the longer branch at p, and we would get repeating branches of the height greater than 2. Then we may take the left tree of Figure 5.26, cut all the repeating branches off from p and re-attach them to one of the leaves of $\mathcal{A}$. This operation will necessarily produce a different tree since the only edge of the weight 2 will be now at distance 1 from the leaf while it was at distance 2 in the initial tree. Therefore, this case is impossible.

Proposition 5.22 is proved. □

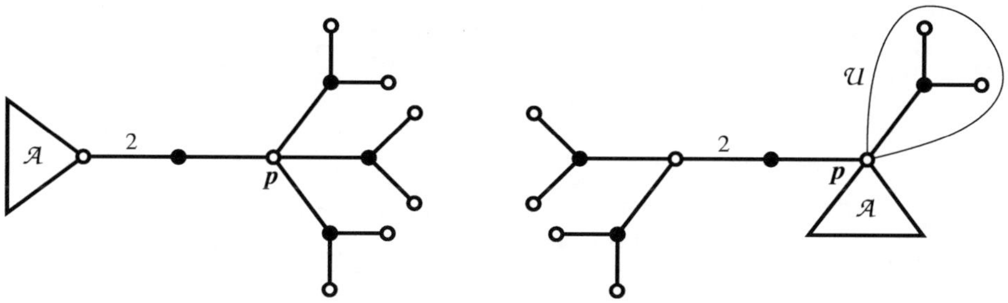

FIGURE 5.26. Illustration to the proof of Proposition 5.22.

5.5. Trees with repeating branches of the type $(1, s, s + 1)$

If a unitree has a crossroad from which grow repeating branches of height > 2, then these branches "start" either with a path $1, s, s + 1$, or with a path $1, t, 1$, where s and t here may be equal to either 1 or 2: see Figures 5.28 (page 46) and 5.35 (page 51). In this section we classify unitrees which have no crossroads of the second type. We start with the following lemma.

LEMMA 5.23 (Repeating branches of the type $(1, s, s + 1)$). *If a unitree has a repeating branch of the type* $(1, s, s + 1)$ *then this branch has one of the two forms shown in* Figure 5.27. *Furthermore, in the second case the unitree is necessarily the tree P.*

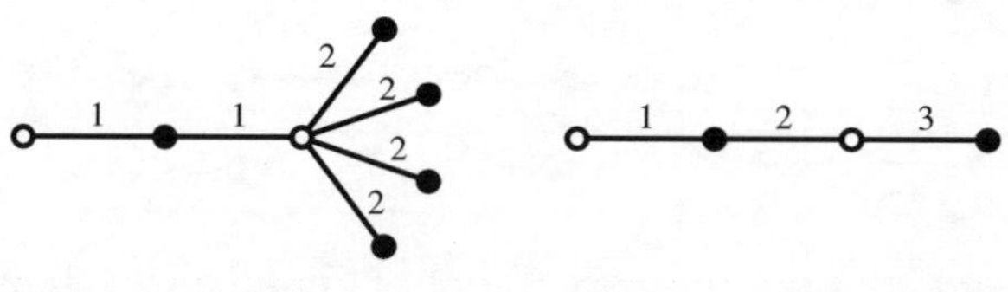

FIGURE 5.27. The only possible forms of repeating branches of the type $(1, s, s + 1)$.

PROOF. First of all observe that the subtrees $\mathcal{A}$ and $\mathcal{C}$ in Figure 5.28 can be interchanged, and if one of them was empty while the other was not, this operation would change the number of leaves, so that the tree in question could not be a unitree. We will show now that the assumption that both trees $\mathcal{A}$ and $\mathcal{C}$ are not empty also leads to a contradiction (so that, in fact, both of them are empty).

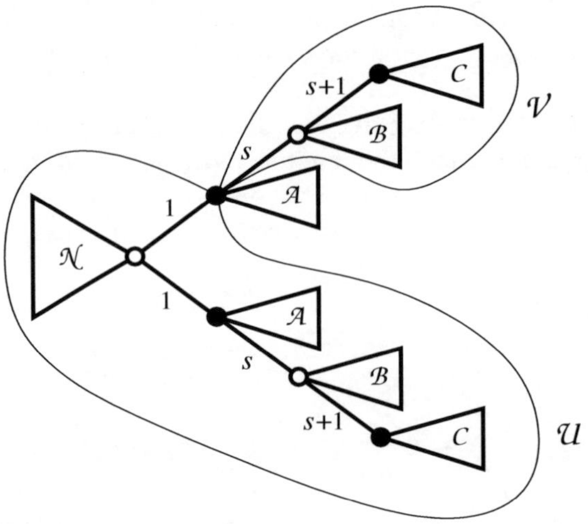

FIGURE 5.28. Illustration to the proof of Lemma 5.23: the subtrees $\mathcal{A}$ and $\mathcal{C}$ are empty; the subtree $\mathcal{B}$ is a bunch of leaves of weight $s+1$.

The tree $\mathcal{V}$ is isomorphic to a subtree of $\mathcal{U}$ and hence is distinct from $\mathcal{U}$. Therefore, if $\mathcal{A}$ is not empty, then it consists of a certain number of copies of $\mathcal{U}$ or of $\mathcal{V}$. The first case is impossible since $\mathcal{A}$ is a subtree of $\mathcal{U}$. Therefore, $\mathcal{A}$ consists of a certain number of copies of $\mathcal{V}$ implying that $\mathcal{C}$ is a proper subtree of $\mathcal{A}$. Now, interchanging $\mathcal{A}$ and $\mathcal{C}$ in every repeating branch we may prove in the same way that $\mathcal{A}$ is a proper subtree of $\mathcal{C}$, implying the contradiction that we need. Thus, $\mathcal{A}$ and $\mathcal{C}$ are empty. In particular, $\mathcal{B}$ is merely a collection of leaves of weight $s+1$.

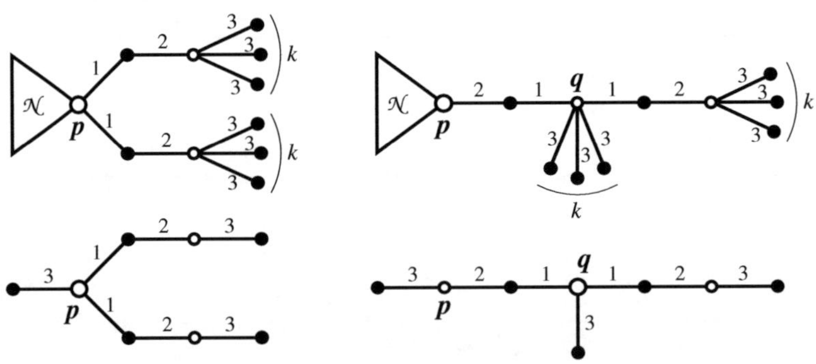

FIGURE 5.29. Illustration to the proof of Lemma 5.23: transformations of repeating branches of height 3 with the weight sequence 1, 2, 3.

Assume now that $s = 2$. Then our tree must look as in Figure 5.29, top left, where the number of repeating branches at the vertex p may be two or more. Let us take two of these branches, apply the transformation shown on top right, and see what takes place at the vertex q. According to Lemma 5.10, all the subtrees growing out of this vertex, except maybe one, must be isomorphic. This can only happen when $k = 1$, and the subtree growing from q to the left is isomorphic to the one growing from q to the right. Therefore, before the transformation there were exactly two (and not more) repeating branches at p, and the subtree $\mathcal{N}$ was reduced to a single leaf of weight 3. The resulting situation is shown in Figure 5.29, bottom left. In this case, our transformation can still be applied, but it leads to a tree isomorphic to the initial one. The unitree thus obtained is P, see Figure 5.9 (page 33).

Lemma 5.23 is proved. □

PROPOSITION 5.24 (Branches of the type $(1, s, s+1)$). *A unitree which has at least one crossroad of type* $(1, s, s+1)$ *but no crossroads of type* $(1, t, 1)$ *belongs to one of the types J, L, M, N, O, P, R, or S.*

PROOF. Once again, this is a long proof: it ends on page 50. In view of Lemma 5.23, if $s = 2$ we get the tree P. Thus, we may now assume that $s = 1$ and the repeating branches have the form shown in Figure 5.27 on the left. Suppose first that the number of the repeating branches is three or more, and apply the transformation shown in Figure 5.30, that is, interchange the positions of a leaf of weight 2 and of a pair of repeating branches.

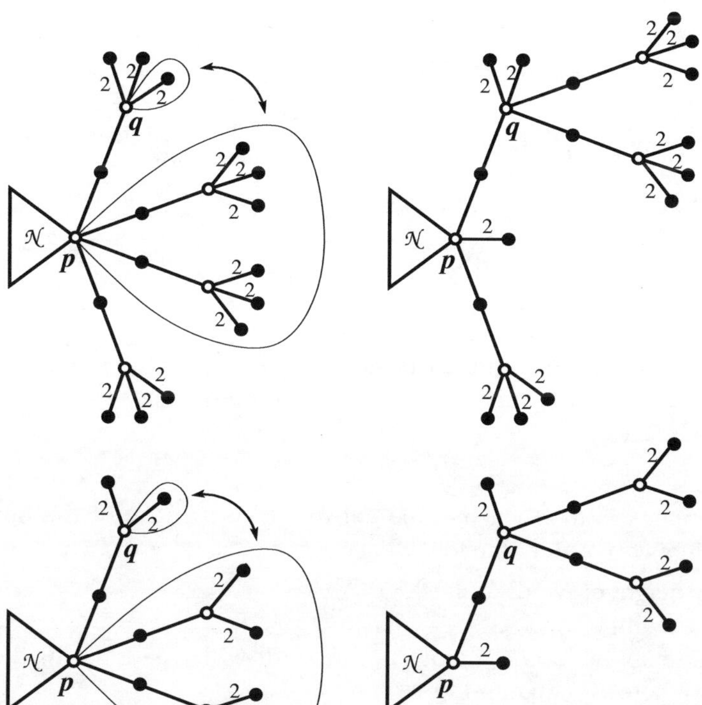

FIGURE 5.30. Illustration to the proof of Proposition 5.24: transformations of repeating branches of height 3 with the weight sequence 1, 1, 2.

If the number of the repeating branches was more than three then the principle "all branches except maybe one are isomorphic" would be violated at the vertex p. The same principle would be violated at the vertex q if the number of leaves in a repeating branch was more than two. Therefore, the number of repeating branches is three, and our transformation looks as is shown in Figure 5.30, bottom. If the number of leaves in a repeating branch is two, then applying once again the same principle at the point q, we arrive at the tree O (Figure 5.9, page 33). Assume now that this number is equal to one. Then the height of $\mathcal{N}$ is less than two, since otherwise the new tree would have more crossroads than the initial one. Furthermore, if $\mathcal{N}$ is a bunch of leaves of weight two, then we could transfer all these leaves to the vertex q thus changing the tree. Therefore, $\mathcal{N}$ is empty, and we arrive at the tree N (see Figure 5.9 once again).

Suppose next that the number of repeating branches is two. The starting edge of the non-repeating branch $\mathcal{N}$ is either of weight 2, and then, according to Proposition 5.18, it is a leaf, and we get the tree J (Figure 5.7, page 33), or it is of weight 1. In the latter case it does not have to be a leaf, though this situation imposes another constraint: the repeating branches must have only one leaf, otherwise the transformation shown in Figure 5.31 can be applied, producing three non-isomorphic branches growing from the crossroad.

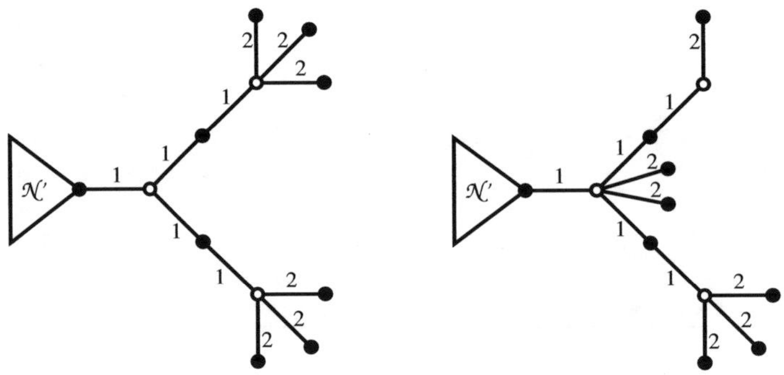

FIGURE 5.31. Illustration to the proof of Proposition 5.24: two repeating branches. What may take place if repeating branches have more than one leaf.

What remains is to study more attentively the structure of the non-repeating branch $\mathcal{N}$. Here we consider the following cases:

- The height of $\mathcal{N}$ is 1, that is, $\mathcal{N}' = \emptyset$.
- The height of $\mathcal{N}$ is 2.
- The height of $\mathcal{N}$ is 3 or more, and $\mathcal{N}$ starts with a path having the weights 1, s, 1 where s is equal to 1 or 2.
- The height of $\mathcal{N}$ is 3 or more, and $\mathcal{N}$ starts with a path having the weights 1, s, $s+1$ where s is equal to 1 or 2.

The height of $\mathcal{N}$ equal to 1 case is trivial: we get the tree L (Figure 5.9).

The height of $\mathcal{N}$ equal to 2 case is illustrated in Figure 5.32. If the weight of the leaves of the non-repeating branch is equal to 2 then, whatever is their number, the reattachment shown on the left changes the tree since the new tree has one leaf less

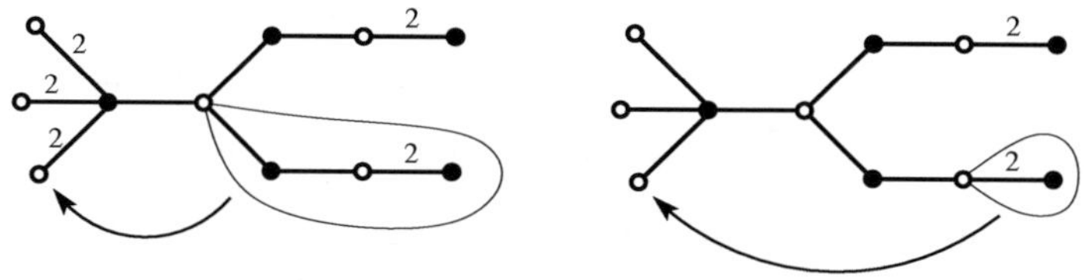

FIGURE 5.32. Illustration to the proof of Proposition 5.24: non-repeating branch of height 2.

than the initial one. If the weight of the leaves of the non-repeating branch is equal to 1 then the reattachment shown on the right also changes the tree unless there is only one leaf in the non-repeating branch. The latter case gives us the tree M (Figure 5.9, page 33).

Assume now that the height of the non-repeating branch is ≥ 3, and this branch contains a path with the weights 1, s, 1, see the upper tree in Figure 5.33.

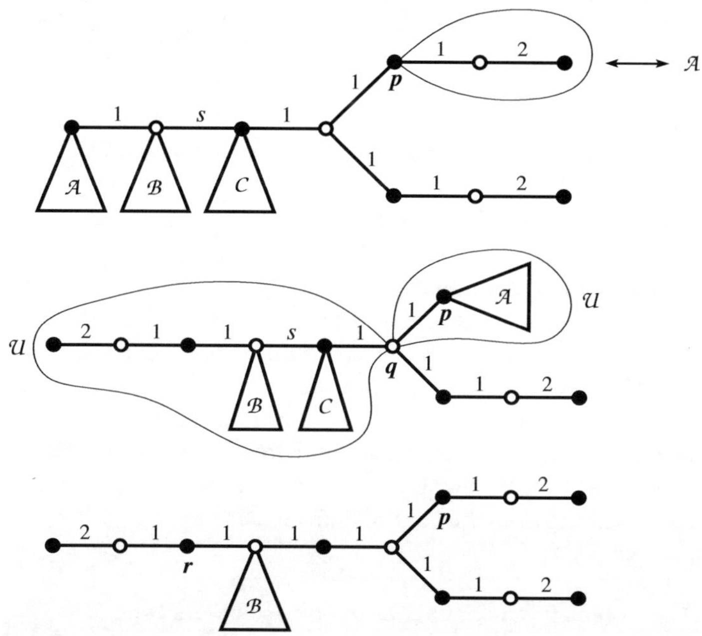

FIGURE 5.33. Illustration to the proof of Proposition 5.24: a non-repeating branch containing a path with the weights 1, s, 1.

First of all, we remark that the subtree $\mathcal{A}$ may be interchanged with a chain of length 2 attached to the vertex p. Then, according to the principle "all branches except maybe one are isomorphic", two situations may occur. First, we could thus create two repeating branches $\mathcal{U}$ attached to the vertex q, see the tree in the middle. But such a tree would contain repeating branches of the type $(1, s, 1)$ which contradicts our supposition. The other possibility is that $\mathcal{A}$ is equal to the chain which was attached to p. Then we get a vertex r (see the lower tree) which is of degree 2 and is incident to two edges of weights 1 and 1. Therefore, according to Proposition 5.18, the edge of weight s, which is not a leaf, cannot have weight 2; hence, $s = 1$. Finally, we affirm that $\mathcal{C} = \emptyset$, otherwise it could be reattached to the vertex p and we would get three different subtrees attached to q. The resulting tree is shown in Figure 5.33, bottom. If $\mathcal{B} = \emptyset$ we get the tree S (Figure 5.10, page 34). If $\mathcal{B} \neq \emptyset$ then, again, according to the principle "all branches except maybe one are isomorphic", $\mathcal{B}$ must be equal to a chain of weights 1, 1, 2, and we get the tree R (Figure 5.10), since otherwise a tree would contain repeating branches of the type $(1, s, 1)$. (Note that in the last case we obtain the tree T which is considered in Proposition 5.25 which treats the case of repeating branches of the type $(1, t, 1)$.)

Finally, consider the case when the non-repeating branch is of height ≥ 3 and contains a path with the weights $1, s, s+1$, see Figure 5.34.

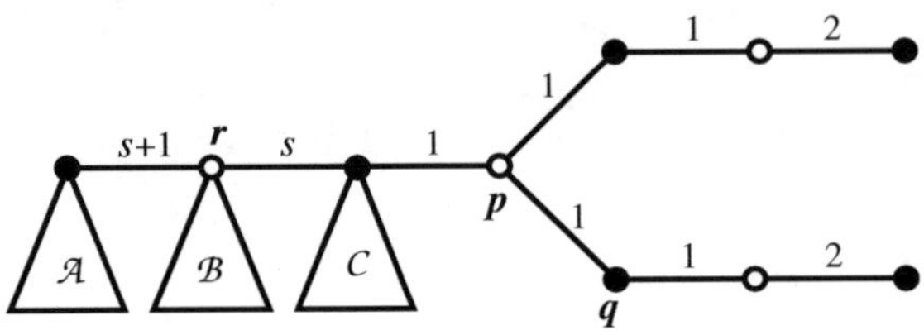

FIGURE 5.34. Illustration to the proof of Proposition 5.24: a non-repeating branch containing a path with the weights $1, s, s+1$.

We affirm that in this case $s = 1$ and all the three subtrees $\mathcal{A}$, $\mathcal{B}$, $\mathcal{C}$ are empty, so that we get the tree N of Figure 5.9, page 33 (and what we call "non-repeating branch" is in this case equal to the repeating branches). Indeed, the tree contains a path with the weights 1, 1, 1; therefore, according to Proposition 5.14, the only possible weights are 1 and 2, so $s = 1$. Now, if $\mathcal{A} \neq \emptyset$ then it could be reattached to the vertex q thus producing a tree with one more leaf. Therefore, $\mathcal{A} = \emptyset$. Then, $\mathcal{C}$ is also empty since $\mathcal{A}$ and $\mathcal{C}$ can be interchanged. Finally, if $\mathcal{B} \neq \emptyset$ then there are two possibilities. Either $\mathcal{B}$ is a bunch of leaves of weight 2; but then it can be reattached to the vertex p. Or $\mathcal{B}$ is a number of copies of the long branch growing out of the vertex r; but then, once again, we would create repeating branches of the type $(1, t, 1)$.

Proposition 5.24 is proved. □

5.6. Trees with repeating branches of the type $(1, t, 1)$

In this section we classify unitrees which have crossroads of the type $(1, t, 1)$.

PROPOSITION 5.25 (Repeating branches of type $(1, t, 1)$). *A unitree which has at least one crossroad with repeating branches of type* $(1, t, 1)$ *belongs to one of the types I, Q, or T.*

PROOF. Consider the first tree of Figure 5.35. We observe that by Lemma 5.10 the subtree $\mathcal{B}$ is a collection of copies of the subtree $\mathcal{U}$, and the subtree $\mathcal{A}$ is a collection of copies of the subtree $\mathcal{V}$. Further, an *sts*-operation, applied to the first tree of Figure 5.35, gives the second tree shown in this figure. This image implies that there are only two repeating branches growing from the vertex p, otherwise the tree would certainly change.

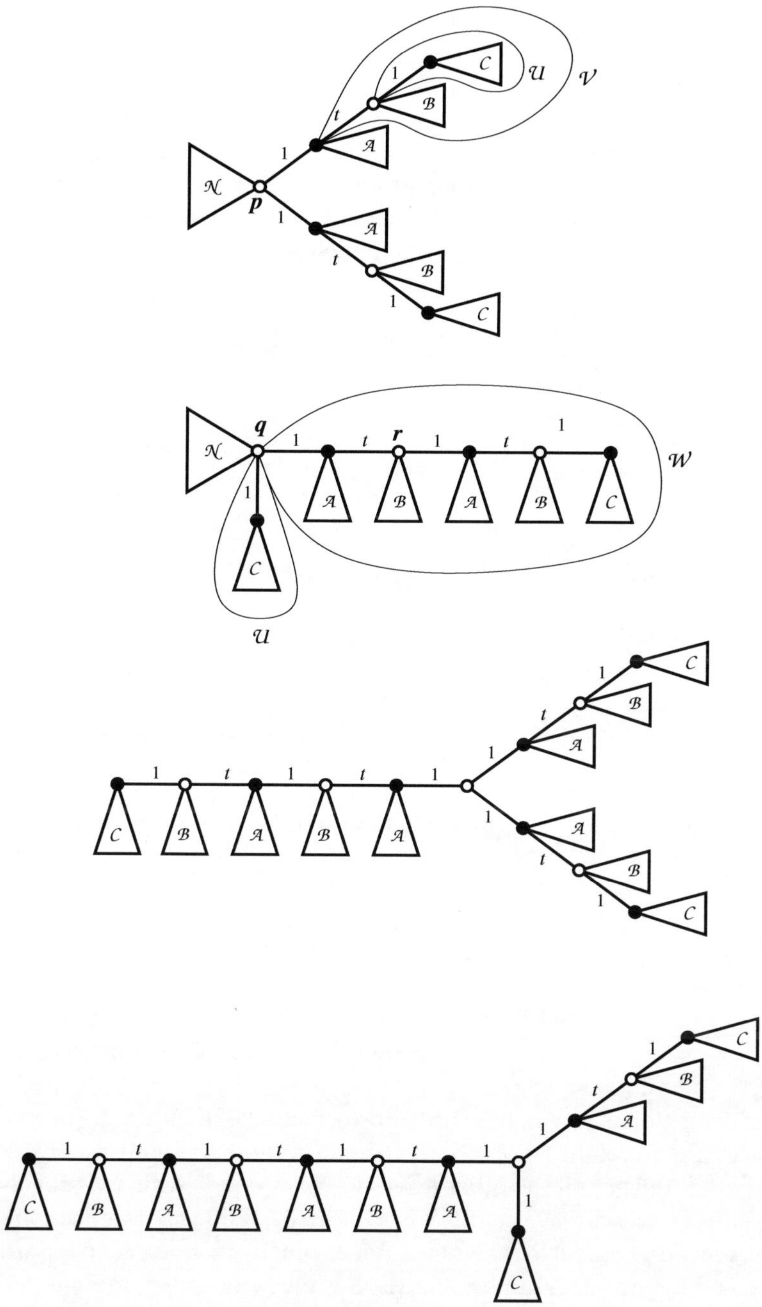

FIGURE 5.35. Illustration to the proof of Proposition 5.25.

Now, looking at the vertex q of the second tree of Figure 5.35 we see that either $\mathcal{N} = \mathcal{U}$ or $\mathcal{N} = \mathcal{W}$. If $\mathcal{N} = \mathcal{W}$ then the initial tree would look like the third tree of the same figure. Then we could once again apply an *sts*-transformation and make the long branch even longer, and one of the repeating branches, shorter (see the fourth tree of the figure), which would give us three different branches attached to p. Hence, $\mathcal{N} = \mathcal{U}$. In particular, we have proved that whenever a unitree has a crossroads of type $(1, t, 1)$ the corresponding non-repeating branch is a subtree of the repeating branch.

Now, it follows from $\mathcal{N} = \mathcal{U}$ that the first tree of Figure 5.35 has a (unique) center at the vertex p while the second one has a (unique) center at the vertex r. Hence, the vertex p of the first tree must correspond to the vertex r of the second one, and thus we must have $t = 1$ and $\mathcal{B} = \mathcal{N} = \mathcal{U}$. Therefore, the tree has the form shown in Figure 5.36, with the same number $l \geq 1$ of branches growing out of the vertices u and v.

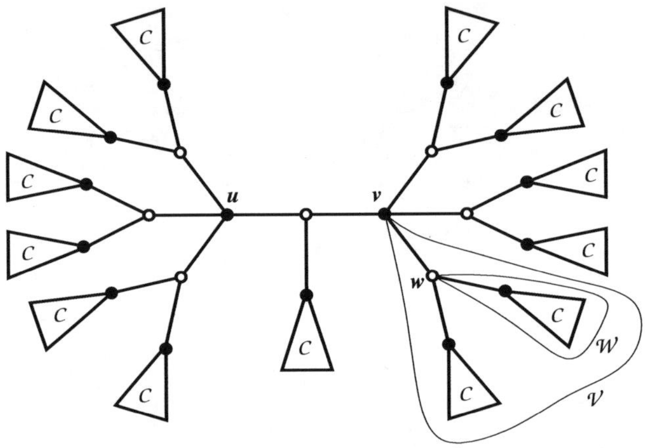

FIGURE 5.36. Illustration to the proof of Proposition 5.25.

If $\mathcal{C} = \emptyset$ then we get a tree of the type I. Assume, next, that $\mathcal{C}$ is non-empty. Observe first that $l = 1$. Indeed, if $l > 1$ then $\mathcal{V}$ is a repeating branch. Furthermore, since $\mathcal{C}$ is non-empty, $\mathcal{V}$ is either $(1, t, t+1)$-branch or $(1, t, 1)$-branch. The first case is impossible by Lemma 5.23, while the second case is impossible since, as we have shown in the previous paragraph, here the corresponding non-repeating branch should be a subtree of $\mathcal{V}$, and this is not so.

If $\mathcal{C}$ is a collection of leaves, then the transformation of Figure 5.22 (page 42) shows that all leaves are of weight 1. Moreover, $\mathcal{C}$ contains no more than one leaf since otherwise we could transfer all the other leaves to the vertex v, which would change the tree. Therefore, in this case we get the tree Q. Finally, if $\mathcal{C}$ is not a collection of leaves, then $\mathcal{W}$ is a repeating branch of height at least 3, which, as above, is necessarily of type $(1, t, t+1)$, and Lemma 5.23 implies that $\mathcal{W}$ has the form shown in Figure 5.27 (page 45) on the left, where the number of leaves is equal to one since otherwise we could transport all the leaves but one to the vertex w. Therefore, in this case we get the tree T.

Proposition 5.25 is proved. □

Phew! It's over at last. We have proved that the trees which do not belong to the collection from A to T are not unitrees. Now we must prove that these survivors are indeed unitrees.

5.7. The trees $A, B, \ldots, T$ listed in Theorem 5.4 are unitrees

Our main tool will be "cutting and gluing leaves", though these operations will be carried out not with the trees themselves but with their passports; the trees must be kept in mind for an intuitive understanding of the proof. We do not repeat every time that "the same reasoning remains valid if we interchange black and white".

We proceed case by case.

(A) There is only one black vertex (see Figure 5.1, page 31); therefore, all white vertices must be adjacent to it, which means that they are leaves. The uniqueness is evident.

The uniqueness proofs for the cases from (B) to (E) all follow the same lines. If the structure of a passport implies the existence of a leaf of degree, say, s, then the only way to construct a corresponding tree is to glue an edge of the weight s to a vertex of the opposite color of a tree with one edge less. Furthermore, in the initial (bigger) tree this edge can only be attached to a vertex of a degree bigger than s. If the smaller tree is a unitree (and usually it is, by induction), and if there is essentially one way to attach the new edge to it, then the bigger tree is also a unitree. In certain cases more than one way of attaching a new edge may exist, but they all lead to isomorphic trees.

(B) Let us consider, for example, the case of an odd length, and examine not the tree itself but its passport (α, β). For this tree, α and β are the same: $\alpha = \beta = ((s+t)^k, s)$. The passport implies that there are two vertices of degree s, a black one and a white one, while all the other vertices, both black and white, are of degree $s + t$. These latter vertices cannot be leaves since otherwise there should exist vertices of a bigger degree to which such leaves would be attached. A tree must have at least two leaves. We conclude that there are exactly two leaves, and they are vertices of degree s. They are connected to the tree by edges of the weight s. The degree of the vertex to which such a leaf is attached is $s + t$, and its color is opposite to the color of the leaf.

Now let us cut off one of these leaves, for example, the white one. Then we get a tree with one less edge and with the passport $\alpha = ((s+t)^{k-1}, s, t)$ and $\beta = (s+t)^k$. This passport corresponds to the chain tree of smaller (and even) length. We may inductively suppose that this tree is a unitree. The vertex of degree t of this tree is a leaf. Now, we must make an operation which would simultaneously fulfill the following three goals:

- it re-attaches back a white leaf of weight s to the smaller tree;
- it makes the black vertex of degree t in the smaller tree to disappear;
- it makes to appear an additional, kth black vertex of degree $s+t$, to the already existing $k - 1$ ones.

It is clear that the only way to do all that is to attach this white leaf of weight s to the black vertex of degree t. This operation re-creates the initial chain-tree.

The proof for an even length repeats the previous one almost word by word, only the leaves are now of the same color and of degrees s and t. The base of induction is a tree consisting of a single edge, which is obviously unique.

(C) The passport of a tree of the type C is $\alpha = (ks + t, s^l)$, $\beta = (ls + t, s^k)$. We affirm that there exists a leaf of degree s. Indeed, a tree must have at least two leaves, and the vertex of the biggest degree cannot be a leaf. The biggest degree is either $ks + t$, or $ls + t$, or both.

Suppose that we have a black leaf of degree s. Then it has to be attached to the unique white vertex of degree bigger than s, which is the white vertex of degree $ls + t$. Cut this leaf off. We get a smaller tree, with one edge less, with l being replaced by $l - 1$. This tree is a smaller instance of C which we may suppose of being a unitree by induction. (Note that in particular cases it can also be of type B, or even of A, the latter one when l was equal to 1.)

Now we no longer work with the passports but with the trees. We know the smaller tree since it is unique, and we must re-attach the previously cut-off black leaf to a white vertex of this smaller tree. Here two cases may take place.

(1) If $s \neq t$, or even if $s = t = 1$ but $l \neq 1$, the initial (bigger) tree did not have a white vertex of degree $2s$. Therefore, we cannot attach the cut-off leaf to a white vertex of degree s. Hence, the only vertex to which it can be attached is the white vertex of degree $(l - 1)s + t$.

(2) If $s = t = 1$ and $l = 1$ then the smaller tree is the star with all its leaves being of degree 1. Then we may re-attach the leaf to any one of them, the resulting tree will be the same.

There is an additional subtlety here. The planar structure of our trees means that we must choose not only a vertex to which we attach a new edge. We must also choose an angle between neighboring edges incident to the vertex of attachment, and to insert the leaf into the angle between these edges. If there are m edges incident to a vertex, there also are m angles between them, and therefore m ways of placing the new edge. But, obviously, in our case all these ways give the same plane tree, see Figure 5.3, page 31.

(D) Two black vertices of degree $2s + t$ cannot be leaves. Therefore, there exists a leaf of degree s or $s + t$. Cut it off, and we get either C or E_1. Indeed, if the cut-off (white) leaf was of degree s, and was attached to one of the black vertices of degree $2s+t$ (there are no other black vertices), then the passport of the smaller tree becomes $\alpha = (2s + t, s + t)$, $\beta = ((s + t)^2, s)$. This passport corresponds to the pattern E_1, with $l = 1$ and the length of the chain equal to 4. The uniqueness of the corresponding tree will be proved in a moment. The only way to glue a leaf of degree s to this tree and to create a vertex of degree $2s + t$ instead of a vertex of degree $s + t$ is to glue this leaf to the vertex of degree $s + t$.

If, on the other hand, the cut-off (white) leaf was of degree $s+t$, and was attached to one of the black vertices of degree $2s + t$, then the passport of the smaller tree becomes $\alpha = (2s + t, s)$, $\beta = (s + t, s^2)$. This passport corresponds to a tree of type C, with $k = 2$ and $l = 1$. The uniqueness of such a tree was proved above. Then the only way to glue back the leaf of degree $s + t$ is to glue it to the black vertex of degree s of the smaller tree.

It is easy to see that in both cases we get the same tree D.

(E) The proof is similar to the cases considered above, so we will shorten our presentation. Consider first the cases E_3 and E_4. All the vertices except two are of degree $s + t$; the two remaining ones are of degrees $(k + 1)s + kt$ and $(l + 1)s + lt$ for E_3, and $(k + 1)s + kt$ and $ls + (l + 1)t$ for E_4. Without loss of generality we may suppose that $(k + 1)s + kt$ is the bigger of the two; therefore, it cannot be a leaf. For E_4, the "second best" vertex cannot be a leaf either since it has the same color. For E_3, if $k > l$, the vertex of degree $(l+1)s + lt$ might in principle be a leaf. Whatever is the case, there exists a leaf of degree $s + t$. Cut it off, and we obtain a smaller tree, with the possible pattern transitions as follows: $E_4 \to E_4$; $E_3 \to E_3$; $E_4 \to E_2$; or $E_3 \to E_1$, the latter two maybe with renaming the variables.

Now, for the cases E_1 and E_2 the situation is similar. All the vertices except two are of degree $s+t$. The vertex of the biggest degree cannot be a leaf. Therefore, there exists a leaf of degree s or $s+t$. Cut it off, and we get a smaller tree, with the possible pattern transitions as follows: $E_1 \to E_1$; $E_1 \to E_2$; $E_2 \to E_1$; $E_2 \to E_2$, or we may arrive to the patterns A or B.

What remains now is to see that there is only one way to re-attach the cut-off leaf to the smaller unitree.

(F, H, I, Q) The trees F, H, I, Q are ordinary; therefore, the enumerative formula (3.5) (page 24) can be applied.

If $m \neq l$, a tree of the series F is asymmetric, and therefore its contribution to (3.5) is 1. Now, formula (3.5) in this case gives 1; therefore, there is no other tree with this passport.

When $m = l$, a tree of the series F is symmetric, with the rotational symmetry of order k. Therefore, its contribution to (3.5) is $1/k$. Now, the formula itself gives $1/k$; therefore, there is no other tree in this case either.

For the trees of the series H, formula (3.5) gives 1 when $k \neq l$, and gives $1/2$ when $k = l$. This corresponds to the symmetry order of these trees: they are asymmetric when $k \neq l$, and symmetric of order 2 when $k = l$.

The trees of the series I are asymmetric, and formula (3.5) gives 1.

The tree Q is asymmetric, and formula (3.5) gives 1.

(G) The tree has km vertices and hence $km - 1$ edges. Since the total weight is km, there exists exactly one edge of weight 2 while all the other edges are of weight 1. The only white vertex to which the edge of weight 2 can be attached is the vertex of degree k, since all the other white vertices are of degree 1. The rest is obvious.

(J) All white vertices are of degree 2; therefore, a weight of an edge can only be 1 or 2. There are only three black vertices, their degrees being 4, $2k+1$, $2k+1$. Therefore, the black vertices cannot be leaves since such leaves could not be attached to a white vertex of degree 2; thus, all the leaves are white. A white vertex which is not a leaf must have two black neighbors; therefore, there are exactly two white vertices which are not leaves: they are "intermediate" white vertices between the black ones. The edges incident to them are both of weight 1. The black vertex of degree 4 cannot have two incident edges of weight 2 since these edges should be leaves, and such a tree would not be connected; it cannot have four incident edges of weight 1 either since such a tree should need more than three black vertices. Therefore, the weights of the edges attached to this vertex must be 2, 1, 1, and the edge of weight 2 is a leaf. The rest is obvious.

(P) All black vertices are of degree 5, all white ones are of degree 3. Therefore, black vertices cannot be leaves. Arguing now as in the case (J) we conclude that there are exactly three white leaves, implying easily that there exists only one tree with this passport.

(K, L, M, N, O, R, S, T) The proof of all these cases follows the same pattern.

Let us take, for example, the tree O. Its passport is $(5^4, 2^{10})$. Therefore, the number of vertices is $4 + 10 = 14$ and the number of edges is 13, while the total weight is $5 \cdot 4 = 2 \cdot 10 = 20$. Thus, the overweight is 7, and it must be distributed among the edges.

Now, no edge can have a weight greater than 2 since the degrees of white vertices are all equal to 2. Therefore, the tree O has seven edges of weight 2. Moreover, *all of them are leaves*; indeed, if something were attached to the white end of such an edge then this white end would have a degree greater than 2.

The same reasoning may be carried out for all the above cases, with their respective overweights and numbers of leaves of weight 2.

Now, let us cut off all the leaves of weight 2. What remains is an ordinary tree, and we must verify that it is a unitree. Usually it is immediately obvious since the ordinary tree in question is very small; otherwise, we may apply formula (3.5), or else we may remark that such a tree belongs to one of the previously established cases. For example, for the tree T what remains after cutting off the leaves of weight 2 is the tree Q.

The last step consists in proving that there is only one way to glue back to this ordinary unitree the leaves of weight 2 that were previously cut off. For example, in the case O the ordinary unitree has black vertices of degrees 3, 1, 1, 1, and, by gluing to them seven edges of weight 2, we must make these degrees equal to 5, 5, 5, 5. Obviously, there is only one way to do that. In fact, in certain cases there are several ways of gluing but they give the same result because of a symmetry of the underlying ordinary tree. For the tree L, there is an additional condition we must satisfy: white leaves can only be attached to black vertices.

Theorem 5.4 is proved! □

5.8. Appendix: the inverse enumeration problem

The usual approach of the enumerative combinatorics is as follows: we consider a class of objects, fix some of their characteristics and try to find the number of objects having these characteristics. For example, we may want to find the number of trees with a given total weight, or with a given number of edges, or even with a given passport.

Inverse enumeration problem. This problem consists of fixing the number, say k, of objects and trying to classify the characteristics and the objects themselves such that there are exactly k of them. In this chapter we have solved the inverse problem for $k = 1$. One can see that it is not an easy exercise, but it is exactly this statement of the problem which is pertinent to the theory of dessins d'enfants.

Below we illustrate by an example what may happen for $k = 2$. Consider the two-parameter family of passports $(r^2, s^3 v^1)$; they depend on r and s while $v = 2r - 3s$ is determined by r and s. Obviously, $3s < 2r$, so that $s < 2r/3$. Three cases are possible: see Figure 5.37. We suppose that the expressions in the figure, like $r - 2s$ or $r - s - v$, are never equal to zero, otherwise some trees may split in two disjoint parts.

Orbit A. This orbit corresponds to the range $r/2 < s < 2r/3$. Indeed, the weight of one of the edges is $r - s - v$. In order for it to be positive we need $r - s - (2r - 3s) = -r + 2s > 0$, so that $s > r/2$. The degree of the white vertex which is the third from the left is equal to $(r - s) + (r - s - v) = 2r - 2s - (2r - 3s) = s$.

The trees are symmetric to each other with respect to the horizontal axis. Hence they are defined over an imaginary quadratic field, and the complex conjugation transposes one of them into the other. A simple computation shows that the quadratic field for all the three orbits is $\mathbb{Q}(\sqrt{\Delta})$ where $\Delta = 3(r - 2s)(2r - 3s)$. When $r/2 < s < 2r/3$ we have indeed $\Delta < 0$. For the other two orbits $\Delta > 0$, so that both times the field is real, and in the majority of cases, quadratic.

Orbit B. This orbit corresponds to the range $r/3 < s < r/2$. Indeed, there are edges of weights $r - 2s$ and $r - v$. In order to have $r - 2s > 0$ we need $s < r/2$, and in order to have $r - v > 0$ we need $r - (2r - 3s) = -r + 3s > 0$ so that $s > r/3$.

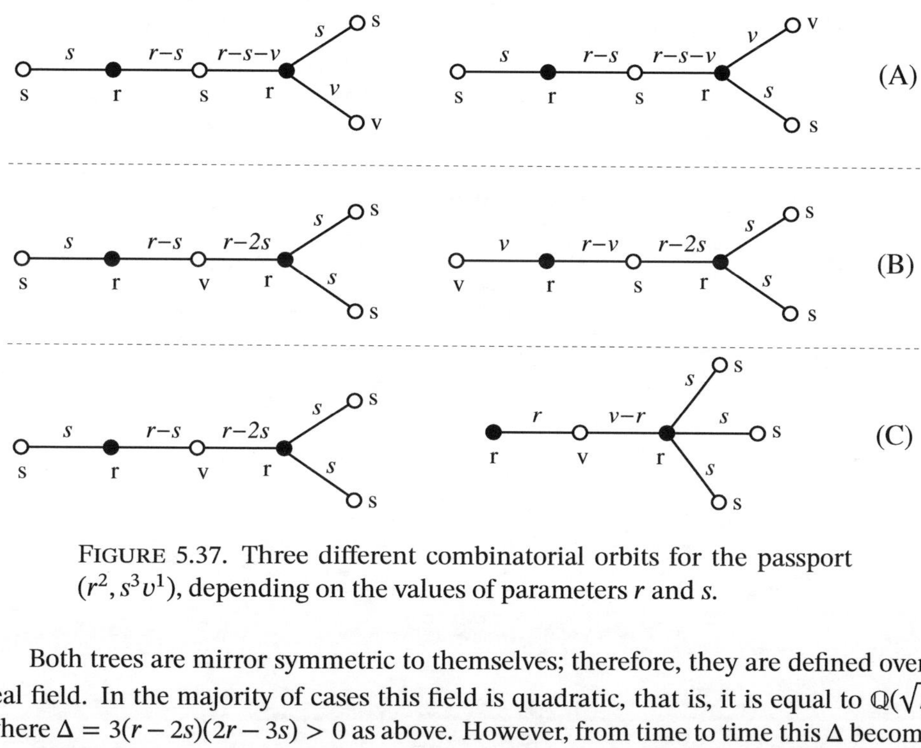

FIGURE 5.37. Three different combinatorial orbits for the passport $(r^2, s^3 v^1)$, depending on the values of parameters r and s.

Both trees are mirror symmetric to themselves; therefore, they are defined over a real field. In the majority of cases this field is quadratic, that is, it is equal to $\mathbb{Q}(\sqrt{\Delta})$ where $\Delta = 3(r - 2s)(2r - 3s) > 0$ as above. However, from time to time this Δ becomes a perfect square. In this case the combinatorial orbit splits into two Galois orbits, so that both trees become defined over $\mathbb{Q}$. We will not treat this case in detail here since in Chapter 10 we will consider a much more beautiful example of this phenomenon.

Notice that the orbit A cannot split into two. Indeed, the complex conjugation is an element of the absolute Galois group (in fact, the only element of this group known explicitly), and its action on both trees is non-trivial.

Orbit C. This orbit corresponds to the range $s < r/3$. Indeed, the right-hand tree has an edge of the weight $v - r$, and in order for it to be positive we need $v - r = (2r - 3s) - r = r - 3s > 0$, so that $s < r/3$. The degree of the black vertex, third from the left, is $(v - r) + 3s = (2r - 3s - r) + 3s = r$.

It is funny to observe that the left-hand tree is structurally the same as the left-hand tree of the orbit B. However, the values of the parameters have changed, and because of that this tree has changed its companion.

What is a series? We have encountered, for weighted trees, two phenomena which do not manifest themselves in the world of ordinary trees. First, in the same series we get both real and imaginary fields. And, second, along with changing a parameter, a tree also changes its companion. Our colleague George Shabat often asks the following question: *what is a series?* We work, day in and day out, with dessins agglomerated in parametric series but still don't have a formal definition of what a series is. By the way, in the above example the algebraic expression of the Belyĭ functions is the same for all the three cases.

Factorization of discriminants. The fact that the orbit contains two trees implies that, in the generic case, the moduli field is the splitting field of a quadratic polynomial. Can we say something about the discriminant of this polynomial without computations, by merely looking at the pictures? In fact, yes. For example, when the value of the parameter s passes from right to left through the threshold $r/2$, the field, which was imaginary, becomes real. This means that the discriminant changes its sign. Therefore, it must contain a factor $r - 2s$.

Does there exist another useful information concerning the discriminant that could be deduced from the pictures? For the time being, this path remains entirely unexplored.

CHAPTER 6

Computation of Davenport–Zannier pairs for unitrees

It is time to meditate a little on what we have achieved and, especially, *how* have we achieved it.

In the previous chapter, we classified all unitrees. The prefix "uni" means that the combinatorial orbit of such a tree consists of a single element. We also know that a Galois orbit is always a subset of a combinatorial orbit; thus, the Galois orbit of a unitree also consists of a single element. The latter means that the absolute Galois group $\mathrm{Aut}(\overline{\mathbb{Q}}|\mathbb{Q})$ acts on such a tree trivially. And this can only happen when the tree in question is defined over $\mathbb{Q}$ (since $\mathbb{Q}$ is the only number field on which the absolute Galois group acts trivially).

What is the conclusion? The conclusion is as follows. We take a unitree; we right down a system of algebraic equations satisfied by the coefficients of the corresponding DZ-polynomials. The system may be extremely complicated, and we may even be unable to solve it. What we *may* do is to affirm that this system has a solution in rational numbers. This is a very strong statement, isn't it? Strong and *concrete*.

And how did we arrive to this conclusion? By two ways. First, by some very abstract considerations of the two above paragraphs concerning the action of the absolute Galois group. And, second, by playing around with trees: cutting edges, reattaching subtrees, etc.

Mathematics is indeed a very amazing edifice. A long-range interaction between its remote parts often looks as a sheer miracle.

The goal of this chapter is to compute DZ-pairs corresponding to the unitrees. The exposition is based on the paper [**PaZv-18**] by Pakovich and Zvonkin.

6.1. Reciprocal polynomials

It turns out that technically it is often much more convenient to work not with the polynomials appearing in DZ-pairs but with their *reciprocals*. Coefficients of a polynomial reciprocal to P go in the opposite direction as compared to P itself.

DEFINITION 6.1 (Reciprocal polynomial). For a polynomial P of degree n, its *reciprocal* is $P^*(x) = x^n \cdot P(1/x)$.

In many examples, the reciprocals of polynomials forming a DZ-pair take the form of initial segments of power series of certain special functions. After having observed this phenomenon we learned that it was (re)discovered many times, notably in [**Dan-89**], [**Adr-97b**], [**BeSt-10**].

Assume that polynomials P and Q of degree n form a DZ-pair, so that

$$\deg(P - Q) = (n+1) - (p+q) = n - (p+q-1), \tag{6.1}$$

and denote by m the number of edges of the corresponding topological tree. This tree has $p+q$ vertices, therefore it has $m = p+q-1$ edges. Considering P and Q as power series we may write condition (6.1) as

$$P - Q = O(x^{n-m}) \quad \text{when} \quad x \to \infty. \tag{6.2}$$

For the reciprocal polynomials condition (6.1) is transformed into the following one:

$$P^* - Q^* = x^m \cdot S, \tag{6.3}$$

where S is a polynomial. Equivalently, we may write

$$P^* - Q^* = O(x^m) \quad \text{when} \quad x \to 0. \tag{6.4}$$

For instance, in Example 1.2 (page 1) the polynomials reciprocal to P and Q and to their difference look as follows:

$$\begin{aligned}
P^* &= (1 - 2x + 33x^2 - 12x^3 + 378x^4 + 336x^5 + 2862x^6 \\
&\quad + 2652x^7 + 14\,397x^8 + 9922x^9 + 18\,553x^{10})^3, \\
Q^* &= (1 - 3x + 51x^2 - 67x^3 + 969x^4 + 33x^5 + 10\,963x^6 \\
&\quad + 9729x^7 + 96\,507x^8 + 108\,631x^9 + 580\,785x^{10} + 700\,503x^{11} \\
&\quad + 2\,102\,099x^{12} + 1\,877\,667x^{13} + 3\,904\,161x^{14} + 1\,164\,691x^{15})^2, \\
P^* - Q^* &= x^{24} \times 2^6 \cdot 3^{15}(5 - 6x + 111x^2 + 64x^3 + 795x^4 + 1254x^5 + 5477x^6).
\end{aligned}$$

By abuse of notation, we may denote the latter polynomial by R^*. This notation may be interpreted as follows: the polynomial R is considered not as a polynomial of degree 6 but as a polynomial of degree 30 (like P and Q themselves) whose coefficients of degrees from 7 to 30 are equal to zero.

6.2. Remarks about computation

The computation of Belyĭ functions has recently become a vast domain of research. A remarkable overview of this activity may be found in [**SiVo-14**], a paper of 57 pages, with a bibliography of 176 titles. Beside a direct approach, involving the solution of a system of polynomial equations, the authors of [**SiVo-14**] also discuss complex analytic methods, methods envolving modular forms, and p-adic methods.

Taking into account the above considerations, we would like to underline one aspect of our work: though we do compute Belyĭ functions for certain individual dessins, the most interesting, and the most difficult part is the computation of Belyĭ functions for *infinite series* of dessins which depend on one or several parameters. As we have seen, the computation for an individual dessin may be very difficult (see, for example, the discussion on page 2). However, for an infinite series the situation is significantly more complicated. Usually, the first thing to do is to compute quite a few particular cases, sometimes dozens of them (or to use other heuristics whenever possible). Then, we need to guess a general pattern of the corresponding Belyĭ functions. And, finally, instead of a trivial verification step which was applicable to individual dessins, we should provide a *proof*, which may turn out to be rather laborious.

In this chapter we obviously do not expose the first step of the above procedure. What we do is a presentation of the final results, that is, of the general form of Belyĭ functions in question, and then we give the proofs whenever they are necessary.

* * *

As it was already said, the unitrees comprise ten infinite series, from A to J, and ten sporadic trees, from K to T. In the subsequent sections we do not strictly follow the "alphabetic" order of trees since we prefer to underline the structural properties of Belyĭ functions in question. Certain Belyĭ functions are expressed in terms of Jacobi polynomials; there are others which lead to interesting differential relations; we will also encounter compositions, Padé approximants, an application to the Hall conjecture, etc.

6.3. Stars and binomial series

Our first series, called "series A" in Chapter 5, is composed of star-trees, see Figure 6.1. All edges except maybe one are of the same weight. This is a three-parametric series: the parameters are s, t, and the number k of the edges of weight s. The total weight of the tree is $n = ks + t$.

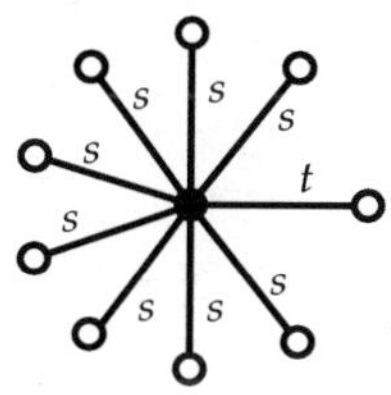

FIGURE 6.1. Star-trees. There are k edges of weight s and one edge of weight t, and $\gcd(s,t) = 1$.

Clearly, we may put the only black vertex at $x = 0$, put the white vertex of degree t at $x = 1$, and assume that both P and Q are monic. Then $P(x) = x^n$ and

$$Q(x) = (x-1)^t \cdot A(x)^s, \tag{6.5}$$

where A is a monic polynomial of degree k whose roots are the white vertices of degree s. Now, condition (6.2) takes the form

$$x^n - (x-1)^t \cdot A^s \underset{x\to\infty}{=} O(x^{n-(k+1)}). \tag{6.6}$$

The only thing we need to know is the polynomial A.

PROPOSITION 6.2 (Star-trees and binomial series). *The polynomial A^* reciprocal to A is the initial segment of the binomial series for $(1-x)^{-t/s}$ up to the degree k:*

$$(1-x)^{-t/s} \underset{x\to 0}{=} A^* + O(x^{k+1}). \tag{6.7}$$

PROOF. Let us pass to reciprocals in (6.6): we need to obtain A^* such that

$$1 - (1-x)^t \cdot (A^*)^s \underset{x\to 0}{=} O(x^{k+1}).$$

We must verify that the polynomial A^* defined in (6.7) satisfies the latter equality. We have:

$$A^* = (1-x)^{-t/s} + h \cdot x^{k+1}, \tag{6.8}$$

where

$$h \underset{x\to 0}{=} O(1).$$

Therefore,

$$A^*(1-x)^{t/s} = 1 + h\cdot x^{k+1}(1-x)^{t/s},$$

and

$$(A^*)^s(1-x)^t = \left[1 + h\cdot x^{k+1}(1-x)^{t/s}\right]^s \underset{x\to 0}{=} 1 + O(x^{k+1})$$

which concludes the proof. □

6.4. Forks and Hall's conjecture

The two-parametric series of trees shown in Figure 6.2 was called "series D" in Chapter 5.

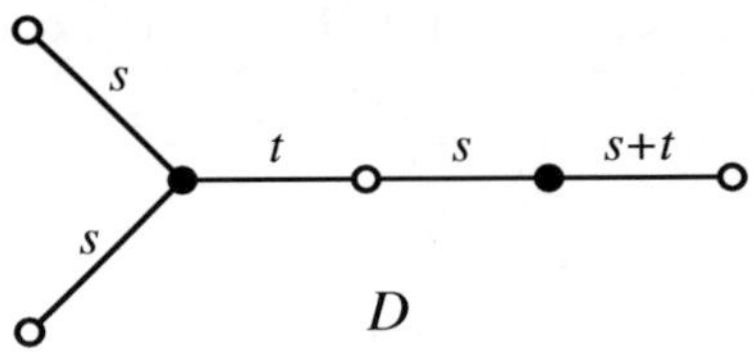

FIGURE 6.2. Fork-trees. There are exactly two leaves of weight s and exactly one leaf of weigh $s+t$. As usual, $\gcd(s,t) = 1$.

6.4.1. Calculation of DZ-pairs. This is the only infinite series of unitrees for which we were able to find the corresponding DZ-pairs directly by a computer calculation. Let us introduce the following three *quadratic* polynomials:

A	–	the roots of A are two black vertices of degree $2s+t$;
B	–	the roots of B are two white vertices of degree $s+t$;
C	–	the roots of C are two white vertices of degree s.

PROPOSITION 6.3 (DZ-pair, fork trees). *We have* $P = A^{2s+t}$ *and* $Q = B^{s+t}\cdot C^s$, *where*

$$A = x^2 - (3s+t)(3s+2t); \tag{6.9}$$

$$B = x^2 - 6s\cdot x + (3s-2t)(3s+t); \tag{6.10}$$

$$C = x^2 + 6(s+t)\cdot x + (3s+2t)(3s+5t). \tag{6.11}$$

PROOF. By (6.4), we must prove that

$$(A^*)^{2s+t} - (B^*)^{s+t}\cdot(C^*)^s = O(x^5). \tag{6.12}$$

Clearly, we may assume that the sum of the roots of A equals zero. Write

$$A^* = 1 - ax^2, \qquad B^* = 1 - bx + cx^2, \qquad C^* = 1 + dx + ex^2,$$

and calculate, with the help of Maple, the first five coefficients of the Taylor series in the left-hand side of (6.12). Equate now the expressions thus obtained to zero and solve the corresponding system in the unknowns a, b, c, d, e. Maple returns two solutions:

$$a = -e, \quad b = 0, \quad c = e, \quad d = 0, \quad e = e,$$

and

$$a = \frac{b^2\left(9\,s^2 + 9\,ts + 2\,t^2\right)}{36s^2}, \qquad b = b, \qquad c = \frac{b^2\left(9\,s^2 - 3\,ts - 2\,t^2\right)}{36s^2},$$

$$d = \frac{(t+s)\,b}{s}, \qquad e = \frac{b^2\,(9\,s^2 + 10\,t^2 + 21\,ts)}{36s^2}.$$

Rejecting the first solution, for which the roots of A, B and C coincide, and making an additional normalization by setting $b = 6s$, we obtain formulas (6.9), (6.10) and (6.11). □

6.4.2. An application: Danilov's theorem. In 1971, M. Hall, Jr. [**Hal-71**] suggested the following two conjectures:

CONJECTURE 6.4 (Distance between a cube and a square of integers).

(1) There exists a constant k such that for all positive integers a, b, $a^3 \neq b^2$, the inequality
$$|a^3 - b^2| > k \cdot a^{1/2}$$
holds.

(2) The exponent $1/2$ in the above inequality cannot be improved. Namely, for every $\varepsilon > 0$ there exists a constant $K(\varepsilon)$ such that there are infinitely many pairs of integers (a, b) satisfying the inequality
$$|a^3 - b^2| \le K(\varepsilon) \cdot a^{1/2+\varepsilon}.$$

The first conjecture is neither proved nor disproved. However, a general belief is that in order to be true it should be modified as follows: for each $\varepsilon > 0$ there exists a constant $k(\varepsilon)$ such that for all positive integers a, b, $a^3 \neq b^2$, the inequality
$$|a^3 - b^2| > k(\varepsilon) \cdot a^{1/2-\varepsilon}$$
holds. In this form the conjecture is a corollary of the famous *abc*-conjecture (see, e. g., [**Lan-90**], [**BeSt-10**] for further details).

As to the second conjecture, in 1982 Danilov [**Dan-82**] proved its stronger version. His result is interesting for us since in his proof he used, in a slightly different normalization, the above polynomials A, B, C, see (6.9), (6.10), (6.11), with the parameters $s = t = 1$.

PROPOSITION 6.5 (Danilov [**Dan-82**]). *There exists a constant K such that there are infinitely many pairs of integers (a, b) satisfying the inequality*
$$|a^3 - b^2| \le K \cdot a^{1/2}. \tag{6.13}$$

PROOF. Specializing (6.9), (6.10) and (6.11) for $s = t = 1$ and computing the difference $P - Q$ we get
$$(x^2 - 20)^3 - (x^2 - 6x + 4)^2(x^2 + 12x + 40) = 1728x - 8640.$$
Substituting $x = 2z$ and dividing both parts by 8 we get
$$(2z^2 - 10)^3 - (2z^2 - 6z + 2)^2(2z^2 - 12z + 20) = 432z - 1080. \tag{6.14}$$
Let us now consider the factor $2z^2 - 12z + 20 = 2(z-3)^2 + 2$ and try to make it a perfect square; then (6.14) will give us a relatively "small" difference between a cube and a square. To do that we have to solve the Diophantine equation
$$u^2 - 2v^2 = 2, \tag{6.15}$$
where $v = z - 3$.

The last equation is a Pell-*like* equation, that is an equation of the form

$$u^2 - Dv^2 = m,$$

where $D > 0$ is not a perfect square, and $m \in \mathbb{Z}$, $m \neq 0$.

We give more details concerning the Pell and Pell-like equations in Chapter 10. Here we provide only the information we need in order to continue the proof of Danilov's theorem.

For $m = 1$ this equation is a usual Pell equation, and it is well known that any Pell equation has infinitely many integer solutions. Pell-like equations not necessarily have integer solutions. However, if at least one such solution (u_0, v_0) exists, then we can obtain infinitely many solutions (u_n, v_n) using the following recursion:

$$u_n + v_n\sqrt{D} \;=\; (u_{n-1} + v_{n-1}\sqrt{D})(k + l\sqrt{D})$$

where (k, l) is the minimum solution of the equation $k^2 - Dl^2 = 1$ with $k, l > 0$. In our case, we easily find that $(k, l) = (3, 2)$.

Equation (6.15) does have an integer solution $(u_0, v_0) = (2, 1)$. Returning to (6.14), it is easy to verify that for all $z \geq 3$ one has

$$432z - 1080 \;<\; 216\sqrt{2}\cdot(2z^2 - 10)^{1/2},$$

which proves the theorem: there are infinitely many pairs of integers (a, b) satisfying (6.13), with the constant $K = 216\sqrt{2}$. □

Experimental results concerning Hall's conjecture may be found in [**LJB-66**] and [**ElkHall**]. Using other DZ-pairs, Danilov [**Dan-89**] and Beukers and Stewart [**BeSt-10**] obtained results similar to Proposition 6.5 for the differences between integer powers a^m and b^n.

6.5. Jacobi polynomials

6.5.1. Trees of this section. Davenport–Zannier pairs for the series of trees considered in this section are expressed in terms of Jacobi polynomials. The trees in question are constructed as follows. First, we take chain-trees with alternating edge weights $s, t, s, t, \ldots$, see Figure 6.3. We must distinguish chains of odd and even length since in one case both ends are of the same color while in the other case they are of different colors.

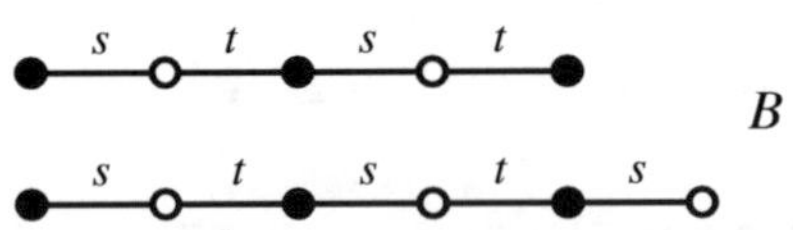

FIGURE 6.3. Series B_1 and B_2: chain-trees.

Then, we have a right to attach to the end-points an arbitrary number of leaves of the weight $s + t$. In this way we obtain "odd" series E_1, E_3 and "even" series E_2, E_4, see Figures 6.4 and 6.5. We call these series "double brushes". Note that any of the parameters k, l, and also both of them, may be equal to zero. Thus, B_1 and E_1 are particular cases of E_3, and B_2 and E_2 are particular cases of E_4.

There are two exceptions from the above construction. The first is when the chain part consists of a single edge, so that there is no alternation of weights. We thus obtain

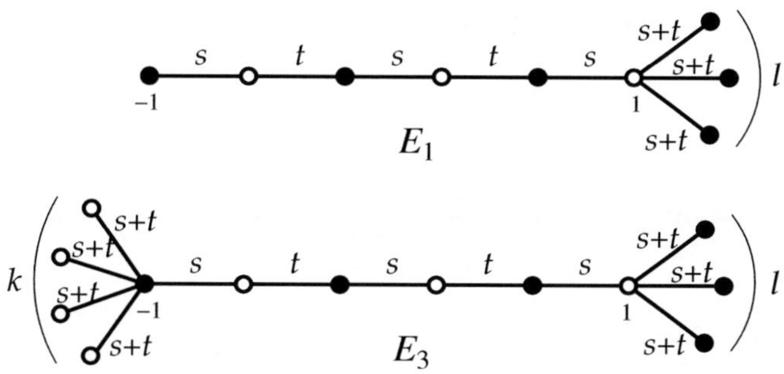

FIGURE 6.4. Series E_1 and E_3: odd double brushes.

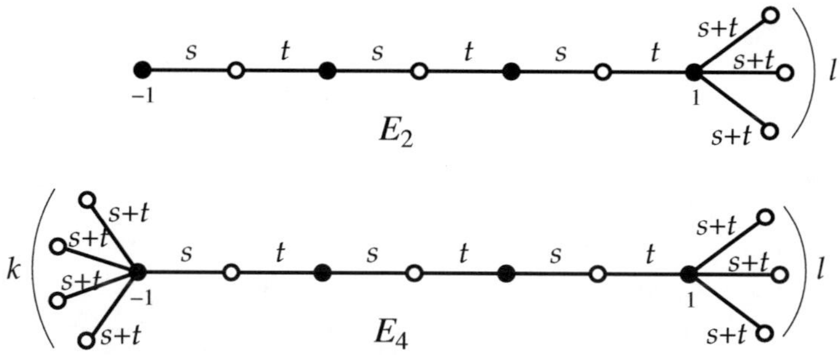

FIGURE 6.5. Series E_2 and E_4: even double brushes.

the series C, see Figure 6.6. In contrast to the general case, now the weight of leaves may be smaller than the weight of the edge between the leaves.

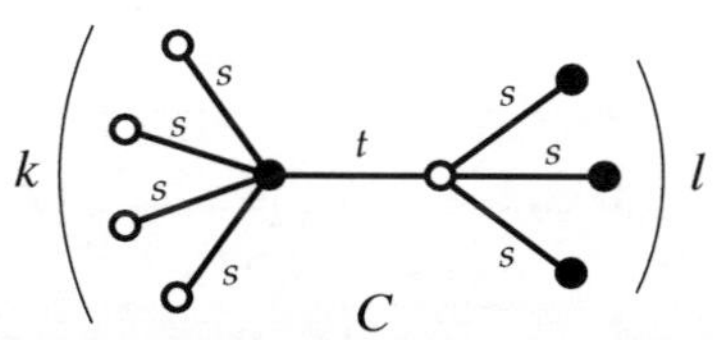

FIGURE 6.6. Series C: trees of diameter 3.

The second exception is when the chain part consists of two edges. In this case it is possible to attach exactly one leaf of weight $s + t$ to one of the ends and exactly two leaves of weight s (or t, to ensure the weight alternation) to the other end. In this way, we get the series of forks D already studied in Section 6.4.

6.5.2. Jacobi polynomials: preliminaries. Let us recall some general facts concerning Jacobi polynomials; for more advanced and detailed treatment see, for example, [**Sze-39**] or [**AbSt-72**].

The classical Jacobi polynomials $J_n(a, b, x)$, $\deg J_n = n$, are defined for the parameters $a, b \in \mathbb{R}$, $a, b > -1$, as orthogonal polynomials with respect to the measure on the segment $[-1, 1]$, given by the density $(1-x)^a(1+x)^b$. The restriction $a, b > -1$ is necessary in order to ensure the integrability of the density. The polynomial $J_n(a,b,x)$ can also be defined as a unique polynomial solution of the differential equation

$$(1-x^2)y'' + [b-a-(a+b+2)x]\,y' + n(n+a+b+1)y = 0, \tag{6.16}$$

satisfying the condition $J_n(a,b,1) = \binom{n+a}{n}$, or by the explicit formula

$$J_n(a,b,x) = \sum_{k=0}^{n} \binom{n+a+b+k}{k}\binom{n+a}{n-k}\left(\frac{x-1}{2}\right)^k. \tag{6.17}$$

Notice that equation (6.16) can be written in the form

$$(1-x^2)Y'' + [a-b+(a+b-2)x]\,Y' + (n+1)(n+a+b)Y = 0, \tag{6.18}$$

where $Y = (1-x)^a(1+x)^b \cdot y$, implying that the function

$$(1-x)^a(1+x)^b \cdot J_n(a,b,x) \tag{6.19}$$

satisfies (6.18).

It follows from (6.17) that $J_n(a,b,x)$ are also polynomials in parameters a and b. Therefore, their definition can be extended to arbitrary (even complex) values of these parameters. These *generalized* Jacobi polynomials still satisfy (6.16), although they are no longer orthogonal with respect to a measure on the segment $[-1,1]$. Similarly, since the function (6.19) may be represented as a power series in x whose coefficients are polynomials in a and b, this function satisfies equation (6.18) for arbitrary a and b.

The following key observation will be used in subsequent proofs. If, in the differential operator (6.16), we replace n with $n+a+b$, replace a with $-a$, and b with $-b$, we get exactly the differential operator (6.18). Therefore, $J_{n+a+b}(-a,-b,x)$ along with (6.19) satisfies (6.18). The last statement, however, should be taken with caution: the subscript $n+a+b$ must be a non-negative integer (hence $a+b$ must also be an integer) since it is the degree of a polynomial.

Notice that if a and b do not satisfy the inequalities $a, b > -1$, then the degree in x of the polynomial $J_n(a,b,x)$ defined by (6.17) may drop down below n. Indeed, (6.17) implies that the leading coefficient of $J_n(a,b,x)$ is equal to

$$\frac{1}{2^n}\binom{2n+a+b}{n} = \frac{1}{2^n \cdot n!}\prod_{i=n+1}^{2n}(a+b+i). \tag{6.20}$$

Hence, in order to obtain a polynomial of degree n we must require that the sum $a+b$ does not take values $-(n+1), -(n+2), \ldots, -2n$. In particular, this is always true if a and b are real and $n \geq -(a+b)$ or, equivalently, $n+a+b \geq 0$.

Along with the density $(1-x)^a(1+x)^b$, which is defined on $[-1,1]$, we will use the multivalued complex function $(z-1)^a(z+1)^b$ (note the change of the sign of the term in the first parenthesis). Clearly, this function has three ramification points $-1, 1, \infty$. Further, observe that if $a+b \in \mathbb{Z}$, then any germ of $(z-1)^a(z+1)^b$ defined near a non-singular point z_0 extends to a function $\mu(z)$ which is single-valued in any domain U obtained from $\mathbb{CP}^1$ by removing a simple curve connecting -1 and 1. Indeed, in such U the function $\mu(z)$ may have a ramification only at infinity. On the other hand, since the analytic continuation of $\mu(z)$ along a loop around infinity is $e^{2\pi(a+b)i}\mu(z)$, we see that ∞ is not a ramification point since $a+b \in \mathbb{Z}$. In particular, $\mu(z)$ can be expanded into a Laurent series at infinity,

$$\mu(z) = c_{a+b}z^{a+b} + c_{a+b-1}z^{a+b-1} + \dots \, .$$

Finally, if a and b are rational numbers, say

$$a = \frac{n_1}{m}, \quad b = \frac{n_2}{m}, \quad n_1, n_2, m \in \mathbb{Z}, \tag{6.21}$$

then any $\mu(z)$ as above satisfies the condition

$$\mu(z)^m = (z-1)^{n_1}(z+1)^{n_2},$$

implying that $\mu(z)$ is defined up to a multiplication by an mth root of unity, and that for a certain choice of this root the equality $c_{a+b} = 1$ holds. By abuse of notation, below we will always use the expression $(z-1)^a(z+1)^b$ to denote the function $\mu(z)$ which satisfies the equality $c_{a+b} = 1$.

LEMMA 6.6 (Jacobi polynomials when $a+b$ is an integer). *Assume that a and b are rational numbers such that $a+b$ is an integer. Then for any $n \geq -(a+b)$ the following equality holds:*

$$\left(\frac{z-1}{2}\right)^a \left(\frac{z+1}{2}\right)^b J_n(a,b,z) - J_{n+a+b}(-a,-b,z) \underset{z\to\infty}{=} O(z^{-(n+1)}). \tag{6.22}$$

PROOF. As it was mentioned above, the function (6.19) satisfies the differential equation (6.18), where the function $\nu(x) = (1-x)^a(1+x)^b$ is assumed to be defined on $[-1,1]$. This function is analytic near the origin; therefore, we can consider its analytic continuation $\nu(z)$, and the function $\nu(z)J_n(a,b,z)$ will satisfy (6.18) in the domain U as above (U is $\mathbb{CP}^1$ with a simple curve connecting -1 and 1 being removed). Furthermore, if (6.21) holds, then

$$\nu(z)^m = (-1)^{n_1}\Big((z-1)^a(z+1)^b\Big)^m,$$

(here n_1 is the numerator of a, see (6.21)) implying that the function

$$(z-1)^a(z+1)^b J_n(a,b,z)$$

also satisfies (6.18) in U.

Since the polynomial $J_n(a,b,x)$ satisfies the differential equation (6.16), we conclude that the functions

$$Y_1 = \left(\frac{z-1}{2}\right)^a \left(\frac{z+1}{2}\right)^b J_n(a,b,z) \quad \text{and} \quad Y_2 = J_{n+a+b}(-a,-b,z)$$

both satisfy the differential equation

$$L_n^{a,b}(Y) = 0 \tag{6.23}$$

where

$$L_n^{a,b} = (1-z^2)\frac{d^2}{dz^2} + [a-b+(a+b-2)z]\,\frac{d}{dz} + (n+1)(n+a+b).$$

This implies that the function $Y_0 = Y_1 - Y_2$ also satisfies this equation. On the other hand, it is easy to see that if $Y(z)$ is a function whose Laurent expansion at infinity is

$$Y = C_d z^d + C_{d-1}z^{d-1} + \dots,$$

then

$$L_n^{a,b}(Y) = \widetilde{C}_d z^d + \widetilde{C}_{d-1} z^{d-1} + \dots$$

where

$$\begin{aligned}\widetilde{C}_d &= -d(d-1) + d(a+b-2) + (n+1)(n+a+b) \\ &= (n+a+b-d)(d+n+1).\end{aligned}$$

Therefore, if Y satisfies (6.23) and $C_d \neq 0$ while $\widetilde{C}_d = 0$, we should have either $d = n+a+b$ or $d = -(n+1)$. Finally, (6.17) implies that the leading terms of both Y_1 and Y_2 are equal to

$$\frac{1}{2^{n+a+b}} \binom{2n+a+b}{n} z^{n+a+b}.$$

Therefore, the degree of the leading term of their difference $Y_0 = Y_1 - Y_2$ is less than $n+a+b$, hence the only possible case is $d = -(n+1)$, implying (6.22). □

6.5.3. Double brushes of even length. Let $\mathcal{T}$ be a weighted tree from the series E_4 or of its two particular cases E_2 or B_2, see Figures 6.5 and 6.3. Denote by r the number of white vertices of $\mathcal{T}$ which are not leaves. Then the total weight of $\mathcal{T}$ is equal to $(s+t)(k+l+r)$ and the total number of edges is equal to $k+l+2r$. Clearly,

$$P = (x-1)^{l(s+t)+t}(x+1)^{k(s+t)+s} \cdot A^{s+t}, \tag{6.24}$$

$$Q = B^{s+t} \tag{6.25}$$

for some polynomials A and B with $\deg A = r-1$, $\deg B = k+l+r$. Furthermore, by (6.2), we must have:

$$P - Q \underset{x\to\infty}{=} O(x^m),$$

where

$$m = (s+t)(k+l+r) - (k+l+2r) = (k+l+r)(s+t-1) - r. \tag{6.26}$$

PROPOSITION 6.7 (DZ-pair, double brushes, even length). *The polynomials P and Q for the double brushes of even length may be represented as follows:*

$$P(x) = \left(\frac{x-1}{2}\right)^{l(s+t)+t} \cdot \left(\frac{x+1}{2}\right)^{k(s+t)+s} \cdot J_{r-1}(a,b,x)^{s+t}, \tag{6.27}$$

where $J_{r-1}(a,b,x)$ is the Jacobi polynomial of degree $r-1$ with parameters

$$a = \frac{l(s+t)+t}{s+t} \quad \textit{and} \quad b = \frac{k(s+t)+s}{s+t}, \tag{6.28}$$

and

$$Q(x) = J_{k+l+r}(-a,-b,x)^{s+t}. \tag{6.29}$$

PROOF. Since the polynomials A and B in (6.24), (6.25) are defined in a unique way up to a multiplication by a scalar factor, it is enough to show that

$$\left(\frac{x-1}{2}\right)^{l(s+t)+t} \left(\frac{x+1}{2}\right)^{k(s+t)+s} J_{r-1}(a,b,x)^{s+t} - J_{k+l+r}(-a,-b,x)^{s+t} \underset{x\to\infty}{=} O(x^m), \tag{6.30}$$

where a and b are given by (6.28), and m, by (6.26).

Let us represent the left-hand side of (6.30) as a product of two factors using the formula

$$u^{s+t} - v^{s+t} = (u-v)(u^{s+t-1} + u^{s+t-2}v + \dots + v^{s+t-1}), \tag{6.31}$$

where

$$u = \left(\frac{x-1}{2}\right)^a \left(\frac{x+1}{2}\right)^b J_{r-1}(a,b,x), \qquad v = J_{k+l+r}(-a,-b,x).$$

It is easy to see that both u and v are $O(x^{k+l+r})$ near infinity. Let us consider the difference $u - v$. Clearly,

$$k + l + r = r - 1 + a + b.$$

Furthermore, since $k, l, r \geq 0$ the following inequality holds:

$$r - 1 \geq -(a + b) = -(k + l + 1)$$

Therefore, by Lemma 6.6, we have:

$$u - v \underset{x\to\infty}{=} O(x^{-r}).$$

On the other hand,

$$u^{s+t-1} + u^{s+t-2}v + \cdots + v^{s+t-1} \underset{x\to\infty}{=} O(x^{(k+l+r)(s+t-1)}).$$

Thus,

$$u^{s+t} - v^{s+t} \underset{x\to\infty}{=} O(x^m)$$

as required. □

REMARK 6.8 (Preceding results). Belyĭ functions for the series E_2 and E_4 with the parameters $s = t = 1$ were first calculated in the Ph. D. thesis of Nicolas Magot in 1997 [**Mag-97**]. A different proof, proposed by Don Zagier, was given in Chapter 2 of [**LaZv-04**]. We used Zagier's proof as a model for the above construction.

6.5.4. Series E_1 and E_3: double brushes of odd length. Let now $\mathcal{T}$ be a weighted tree from the series E_3 or of its two particular cases E_1 and B_1, see Figures 6.4 and 6.3. As above, denote by r the number of white vertices of $\mathcal{T}$ which are not leaves, so that the total weight of $\mathcal{T}$ is $(s + t)(k + l + r) + s$, and the total number of edges is $k + l + 2r + 1$. Now we must find polynomials P and Q such that

$$P = (x+1)^{k(s+t)+s} \cdot A^{s+t}, \tag{6.32}$$

$$Q = (x-1)^{l(s+t)+s} \cdot B^{s+t} \tag{6.33}$$

for some polynomials A and B with $\deg A = l + r$ and $\deg B = k + r$, and

$$P - Q \underset{x\to\infty}{=} O(x^m),$$

where

$$m = (s+t)(k+l+r) + s - (k+l+2r+1) = (k+l+r)(s+t-1) + s - r - 1. \tag{6.34}$$

PROPOSITION 6.9 (DZ-pair, double brushes, odd length). *The polynomials P and Q may be represented as follows:*

$$P(x) = \left(\frac{x+1}{2}\right)^{k(s+t)+s} \cdot J_{l+r}(a,b,x)^{s+t}, \tag{6.35}$$

and

$$Q(x) = \left(\frac{x-1}{2}\right)^{l(s+t)+s} \cdot J_{k+r}(-a,-b,x)^{s+t}. \tag{6.36}$$

where $J_{l+r}(a,b,x)$ and $J_{k+r}(-a,-b,x)$ are the Jacobi polynomials, and the parameters a and b are as follows:

$$a = -\frac{l(s+t)+s}{s+t} \quad \text{and} \quad b = \frac{k(s+t)+s}{s+t}, \tag{6.37}$$

PROOF. We must show that

$$\left(\frac{x+1}{2}\right)^{k(s+t)+s} J_{l+r}(a,b,x)^{s+t} - \left(\frac{x-1}{2}\right)^{l(s+t)+s} J_{k+r}(-a,-b,x)^{s+t} \underset{x\to\infty}{=} O(x^m) \tag{6.38}$$

where a and b are as in (6.37), and m is defined by (6.34).

Equality (6.38) is equivalent to the following one:

$$\left(\frac{x-1}{2}\right)^{-(l(s+t)+s)} \left(\frac{x+1}{2}\right)^{k(s+t)+s} J_{l+r}(a,b,x)^{s+t} - J_{k+r}(-a,-b,x)^{s+t} = O(x^p), \tag{6.39}$$

where

$$p = m - (l(s+t)+s) = (k+r)(s+t-1) - (l+r+1).$$

On the other hand, since

$$k+r = (l+r)+a+b$$

and

$$l+r \geq -(a+b) = l-k,$$

it follows from Lemma 6.6 that

$$\left(\frac{x-1}{2}\right)^a \left(\frac{x+1}{2}\right)^b J_{l+r}(a,b,x) - J_{k+r}(-a,-b,x) = O(x^{-(l+r+1)}),$$

implying in the same way as in Proposition 6.7 that (6.39) holds. □

6.5.5. Series C and B. The series C is a particular case of the series E of odd length corresponding to the case of r equal to zero. In order to adjust the notation (which, in the series E and C, is slightly different) we must set $r = 0$ and change s to t and t to $s-t$ in formulas (6.35)–(6.36), thus obtaining

$$P(x) = \left(\frac{x+1}{2}\right)^{ks+t} \cdot J_l(a,b,x)^s, \tag{6.40}$$

where $J_l(a,b,x)$ is the Jacobi polynomial of degree l with parameters

$$a = -\frac{ls+t}{s} \quad \text{and} \quad b = \frac{ks+t}{s}, \tag{6.41}$$

while

$$Q(x) = \left(\frac{x-1}{2}\right)^{ls+t} \cdot J_k(-a,-b,x)^s. \tag{6.42}$$

Finally, it is clear that the series B_1 and B_2 (chains of odd and even length) are particular cases of the series E_3 and E_4, so that the Davenport–Zannier pairs for B_1 and B_2 are obtained from those for E_3 and E_4, respectively, by setting $k = l = 0$.

6.5.6. Padé approximants. The above results can be interpreted in terms of Padé approximants for the function $(1-x)^a(1+x)^b$. Recall that if

$$f(x) = \sum_{k=0}^{\infty} c_k x^k$$

is a formal power series, then its *Padé approximant* of order $[n/m]$ at zero is a rational function $p_n(x)/q_m(x)$, where $p_n(x)$ is a polynomial of degree $\leq n$ and $q_m(x)$ is a polynomial of degree $\leq m$, such that

$$f(x) - \frac{p_n(x)}{q_m(x)} \underset{x\to 0}{=} O(x^{n+m+1}). \tag{6.43}$$

Defined in this way, Padé approximants do not necessarily exist. However, if an approximant of a given order does exist, it is unique.

Linearizing the problem by requiring that

$$q_m(x)f(x) - p_n(x) \underset{x\to 0}{=} O(x^{n+m+1}) \tag{6.44}$$

we arrive at the notion of a *Padé form* (p_n, q_m) of order $[n/m]$. Being defined by linear equations, Padé forms always exist (in general, (6.44) does not imply (6.43) since $q_m(x)$ may vanish at zero), and the Padé form of a given order is defined in a unique way up to a multiplication by a constant.

Keeping the notation of Section 6.5.3 we may now reformulate the condition for P and Q to be a Davenport–Zannier pair for the series E of even length as follows (a similar result is also true for the series E of odd length):

PROPOSITION 6.10 (Padé forms, even case). *Let polynomials A and B be as in formulas* (6.24) *and* (6.25)*. Then the pair of their reciprocals (A^*, B^*) is the Padé form of order $[r-1/k+l+r]$ for the function $(1-x)^a(1+x)^b$ with parameters*

$$a = \frac{l(s+t)+t}{s+t} \quad \textit{and} \quad b = \frac{k(s+t)+s}{s+t}. \tag{6.45}$$

PROOF. Since the pairs (P, Q) and (A, B) are both defined up to a multiplication by a constant, it is enough to show that

$$(x-1)^{l(s+t)+t}(x+1)^{k(s+t)+s} \cdot A^{s+t} - B^{s+t} \underset{x\to\infty}{=} O(x^p), \tag{6.46}$$

where

$$p = (k+l+r)(s+t-1) - r.$$

By definition of Padé forms we have:

$$(1-x)^a(1+x)^b A^* - B^* \underset{x\to 0}{=} O(x^{k+l+2r}),$$

implying that

$$(1-x)^{l(s+t)+t}(1+x)^{k(s+t)+s} \cdot (A^*)^{s+t} - (B^*)^{s+t} \underset{x\to 0}{=} O(x^{k+l+2r}), \tag{6.47}$$

(here we use formula (6.31) again though now the factors involved are series in non-negative powers of x). Finally, substituting $1/x$ in place of x in (6.47) and multiplying both sides by

$$x^{(k+l+r)(s+t)}$$

we obtain (6.46). □

PROPOSITION 6.11 (Padé forms, odd case). *Let polynomials A and B be as in formulas* (6.32) *and* (6.33). *Then the pair of their reciprocals* (A^*, B^*) *is the Padé form of order* $[l+r/k+r]$ *for the function* $(1-x)^a(1+x)^b$ *with parameters*

$$a = -\frac{l(s+t)+t}{s+t} \qquad and \qquad b = \frac{k(s+t)+s}{s+t}. \tag{6.48}$$

The proof is similar to the previous one, so we omit it. □

REMARK 6.12 (On Padé approximants). From the computational point of view, a great advantage of Padé approximants is due to the fact that the equations describing them are linear. For this reason, the computation of these approximants by Maple is instantaneous. This remains to be an advantage even in the case like ours when the polynomials in question are "known explicitly". For example, one has to use some astute tricks in order to make Maple work with the generalized Jacobi polynomials whose parameters do not satisfy the condition $a, b > -1$.

A vast literature is devoted to the study of Padé approximants for some particular functions, for example, for the exponential function. To our surprise, we did not find any research concerning Padé approximants for the function $(1-x)^a(1+x)^b$. By the way, our Lemma 6.6 can also be reformulated as a result about Padé forms for this function.

6.6. Series F and G: trees of diameter 4

Below we find DZ-pairs for the series F, see Figure 6.8 (page 75), and the series G, see Figure 6.7 (page 73), using their relations to differential equations. For the series F, which consists of ordinary trees, the corresponding formulas are particular cases of the formulas for *Shabat polynomials* for trees of diameter 4, first calculated by Adrianov [**Adr-07b**].

Since the trees from the series F are ordinary, the degree of $R = P - Q$ is zero, that is, $R = c$ for some $c \in \mathbb{C}$. Therefore, in order to describe the corresponding DZ-pair it is enough to find P and c. This is equivalent to the finding of the Shabat polynomial corresponding to the tree. Similarly, for trees from the series G the degree of R is one, and it is technically easier to provide explicit formulas for P and R rather than for P and Q.

We start with the series G.

6.6.1. Series G.

The polynomial P for the series G takes the form

$$P = A(x)^m, \tag{6.49}$$

where A is a polynomial of degree $k-1$ whose roots are the black vertices (all of them are of degree m). Notice that the number of these vertices does not coincide with the degree of the central vertex since we have one "double" edge.

We choose the normalization of P, Q and $R = P - Q$ in the following way:

- $P = A^m$ where A is monic, $\deg A = k-1$;
- the central vertex is placed at $x = 0$, so that $Q = x^k \cdot B$ where B is monic, $\deg B = n - k$; the roots of B are the white vertices distinct from zero;
- $R = c\,(x-1)$; this means that the pole inside the only face of degree 1 is placed at $x = 1$.

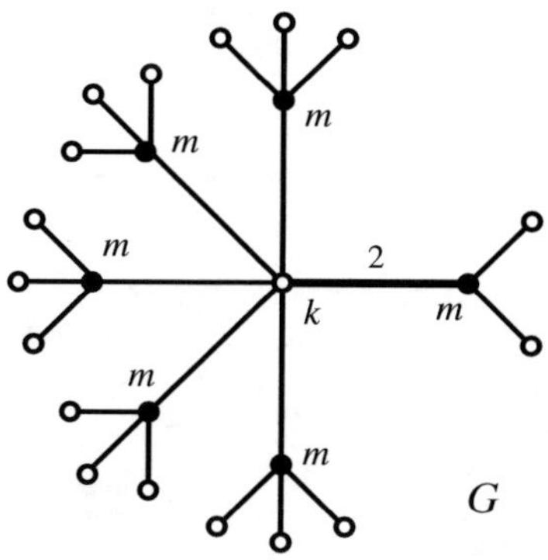

FIGURE 6.7. Series G. The degree of the central vertex is k, the number of branches (and the number of black vertices) is $k-1$.

Thus, we get

$$A^m - c\,(x-1) \;=\; x^k \cdot B. \tag{6.50}$$

PROPOSITION 6.13 (Series G). *The polynomial A whose roots are the black vertices satisfies the differential equation*

$$mA' \cdot (x-1) - A = (m(k-1)-1)x^{k-1}. \tag{6.51}$$

Therefore, the coefficients $a_0, \dots, a_{k-1}$ of $A(x) = \sum_{i=0}^{k-1} a_i x^i$ may be found by the following backward recurrence:

$$a_{k-1} \;=\; 1, \quad a_i \;=\; \frac{m(i+1)}{mi-1} \cdot a_{i+1} \quad \textit{for} \quad 0 \le i \le k-2. \tag{6.52}$$

Finally, $c = -a_0^m$.

PROOF. Taking the derivative of the both sides of equality (6.50) we obtain the equality

$$mA^{m-1}A' - c \;=\; x^{k-1}\,(kB + xB')\,,$$

implying the equality

$$mA^m A' - cA \;=\; x^{k-1}A\,(kB + xB')\,.$$

Substituting in the last equality the expression of A^m from (6.50), we obtain

$$mA'\left[c\,(x-1) + x^k B\right] - cA \;=\; x^{k-1}A\,(kB + xB')$$

and

$$mA' \cdot c\,(x-1) - cA \;=\; x^{k-1}\,[kAB + xAB' - xmA'B]\,.$$

We now observe that the degree of the left-hand side of the last equality is $k-1$, while its right-hand side is *proportional* to x^{k-1}. Therefore, the expression in the square brackets on the right is a constant K, and both parts are equal to $K \cdot x^{k-1}$. The constant K can be easily found as the leading coefficient of the left-hand side: it is equal to $mc(k-1) - c$. Finally, we get the equality

$$mcA' - cA \;=\; (mc(k-1) - c)x^{k-1},$$

which implies (6.51).

Substituting $A(x) = \sum_{i=0}^{k-1} a_i x^i$ in (6.51) we obtain (6.52). Finally, substituting $x = 0$ in (6.50) we obtain $c = -a_0^m$. $\square$

EXAMPLE 6.14 (Series G, $k = 6$). Let us take $k = 6$, so that $\deg A = k - 1 = 5$. Then the corresponding polynomial looks as follows:

$$A = a_5x^5 + a_4x^4 + a_3x^3 + a_2x^2 + a_1x + a_0, \tag{6.53}$$

where

$$\begin{aligned}
a_5 &= 1,\\
a_4 &= \frac{5m}{4m-1},\\
a_3 &= \frac{5m\cdot 4m}{(4m-1)(3m-1)},\\
a_2 &= \frac{5m\cdot 4m\cdot 3m}{(4m-1)(3m-1)(2m-1)},\\
a_1 &= \frac{5m\cdot 4m\cdot 3m\cdot 2m}{(4m-1)(3m-1)(2m-1)(m-1)},\\
a_0 &= \frac{5m\cdot 4m\cdot 3m\cdot 2m\cdot m}{(4m-1)(3m-1)(2m-1)(m-1)(-1)}.
\end{aligned}$$

REMARK 6.15 (Hypergeometric equation). The polynomial A also satisfies the hypergeometric differential equation

$$x(1-x)\frac{d^2y}{dx^2} + \bigl[c-(a+b+1)x\bigr]\frac{dy}{dx} - ab\cdot y = 0. \tag{6.54}$$

Indeed, applying the differential operator $x\dfrac{d}{dx} + (1-k)$ to both parts of the equality (6.51) we obtain

$$x\,[mA'\cdot(x-1)-A]' + (1-k)\,[mA'\cdot(x-1)-A] = 0,$$

implying

$$x(x-1)A'' + \Bigl[\Bigl(1-\frac{1}{m}+(1-k)\Bigr)x-(1-k)\Bigr]A' - \frac{(1-k)}{m}A = 0.$$

Therefore, A is a solution of the differential equation

$$x(1-x)\frac{d^2y}{dx^2} + \Bigl[(1-k)-\Bigl((1-k)-\frac{1}{m}+1\Bigr)x\Bigr]\frac{dy}{dx} + \frac{(1-k)}{m}y = 0$$

which is a particular case of (6.54) with

$$a = 1-k, \quad b = -\frac{1}{m}, \quad c = 1-k.$$

6.6.2. Series F. For this series we may assume that

$$P = (x-1)^l A(x)^m, \qquad Q = x^k B(x). \tag{6.55}$$

Here A is monic and $\deg A = k-1$; namely, A is a polynomial whose roots are the black vertices of degree m. Now, B is a polynomial whose roots are the white vertices distinct from zero, $\deg B = n-k$. The polynomials P and Q must satisfy the condition

$$(x-1)^l A(x)^m - x^k B(x) = c, \tag{6.56}$$

where $c \in \mathbb{C}$ is a non-zero constant.

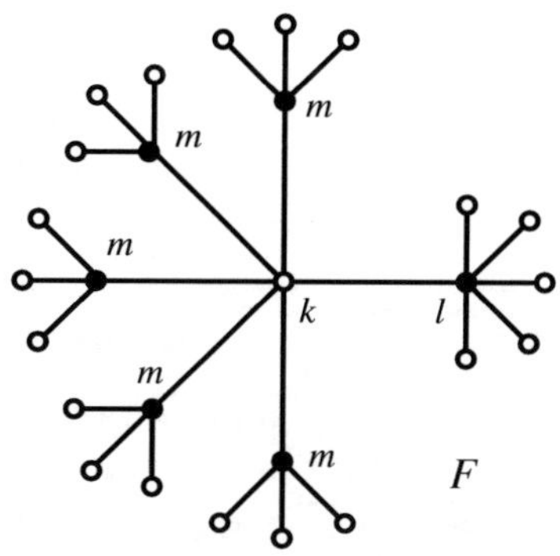

FIGURE 6.8. Series F.

PROPOSITION 6.16 (Series F). *The polynomial A whose roots are the black vertices satisfies the differential equation:*

$$mA' \cdot (x-1) + lA = [m(k-1)+l]\; x^{k-1}. \tag{6.57}$$

Therefore, the coefficients $a_0, \ldots, a_{k-1}$ of $A(x) = \sum_{i=0}^{k-1} a_i x^i$ may be found by the following backward recurrence:

$$a_{k-1} = 1, \quad a_i = \frac{m(i+1)}{mi+l} \cdot a_{i+1} \quad \textit{for} \quad 0 \le i \le k-2. \tag{6.58}$$

Finally, the value of c in (6.56) is equal to $(-1)^l a_0^m$.

PROOF. As above, let us take the derivative of equation (6.56). Then we get

$$(x-1)^{l-1} A^{m-1} \left[lA + m\,(x-1)A'\right] = x^{k-1}\,(kB + xB').$$

We observe that the factor x^{k-1} in the right-hand side is coprime with the factor $(x-1)^{l-1}A^{m-1}$ in the left-hand side, and therefore it must be proportional to the factor $lA + m\,(x-1)A'$ which is itself a polynomial of degree $k-1$. Therefore, the latter polynomial is equal to $K \cdot x^{k-1}$ where the constant K can be found as its leading coefficient; namely, it is equal to $m(k-1)+l$. Thus, (6.57) holds.

Now, substituting $A(x) = \sum_{i=0}^{k-1} a_i x^i$ in (6.57) we obtain the recurrence (6.58), and substituting $x = 0$ in (6.56) we obtain the value of c. □

Here, like in the case of the series G, the polynomial A also satisfies the hypergeometric differential equation, and therefore it may be represented through a hypergeometric function.

EXAMPLE 6.17 (Series F, $k = 6$). Let us take $k = 6$, so that $\deg A = k - 1 = 5$. Then the corresponding polynomial looks as follows:

$$A = a_5 x^5 + a_4 x^4 + a_3 x^3 + a_2 x^2 + a_1 x + a_0$$

where

$$\begin{aligned}
a_5 &= 1\,,\\
a_4 &= \frac{5m}{l+4m}\,,\\
a_3 &= \frac{5m\cdot 4m}{(l+4m)(l+3m)}\,,\\
a_2 &= \frac{5m\cdot 4m\cdot 3m}{(l+4m)(l+3m)(l+2m)}\,,\\
a_1 &= \frac{5m\cdot 4m\cdot 3m\cdot 2m}{(l+4m)(l+3m)(l+2m)(l+m)}\,,\\
a_0 &= \frac{5m\cdot 4m\cdot 3m\cdot 2m\cdot m}{(l+4m)(l+3m)(l+2m)(l+m)l}\,.
\end{aligned}$$

6.6.3. Differential relations. The above method may be applied to DZ-pairs which do not necessary correspond to trees of diameter four or to unitrees. However, in general, it leads to *differential relations* between P and Q. Let us clarify what we mean by considering the problem of the difference between cubes and squares of polynomials.

Let A, B, and R be polynomials such that

$$A^3 - B^2 = R \tag{6.59}$$

and

$$\deg A = 2k, \qquad \deg B = 3k, \qquad \deg R = k+1.$$

Taking the derivative of both parts of (6.59) we obtain

$$3A^2A' - 2BB' = R'.$$

Multiplying now the last equality by A and substituting A^3 from (6.59) we obtain the equality

$$3A'\left(B^2+R\right) - 2BB'A = R'A,$$

implying in its turn the equality

$$B\,(3A'B - 2AB') = R'A - 3A'R.$$

Since the degree of the right-hand side is

$$\deg(R'A - 3A'R) \le 3k$$

while $\deg B = 3k$, the above equality implies that

$$3A'B - 2AB' = c \tag{6.60}$$

for some non-zero constant $c \in \mathbb{C}$.

The last expression is a differential equation of the first order with respect to A as well as with respect to B. Unfortunately, both A and B are unknown. Thus, it does not give us any immediate information about A and B. Still, algebraic equations for coefficients of A and B obtained from (6.60) are (mostly) of degree 2 while the equations obtained from (6.59) are (mostly) of degree 3.

Differentiating (6.60) and writing the expression thus obtained as a differential equation with respect to A we get:

$$A'' + \frac{B'}{3B}\cdot A' - \frac{2B''}{3B}\cdot A = 0. \tag{6.61}$$

This differential equation is a particular case of the differential equation

$$\frac{d^2S}{dz^2} + \left(\sum_{j=1}^{m} \frac{\gamma_j}{z-a_j}\right)\frac{dS}{dz} + \frac{V(z)}{\prod_{j=1}^{m}(z-a_j)} \cdot S = 0, \tag{6.62}$$

where V is a polynomial of degree at most $m-2$. Polynomial solutions of the last equation are called Stieltjes polynomials. The polynomials V for which (6.62) has a polynomial solution are called Van Vleck polynomials. Thus, B is a Van Vleck polynomial, and A is the corresponding Stieltjes polynomial.

Writing now (6.61) in the form

$$B'' - \frac{A'}{2A} \cdot B' - \frac{3A''}{2A} \cdot B = 0$$

we obtain that A is a Van Vleck polynomial and B is the corresponding Stieltjes polynomial.

The above observations show that the relations between DZ-pairs and differential equations may be deeper than it seems at first glance and deserve further investigation.

6.7. Series H and I: decomposable ordinary trees

In this section we consider series H (Figure 6.9) and I (Figure 6.11, page 78). In both cases the corresponding DZ-pairs are obtained using the operation of composition. Notice that the trees in question are ordinary (the weights of all edges are equal to 1). As it was mentioned in Definition 2.22 (page 17), Belyĭ functions for ordinary trees are called Shabat polynomials.

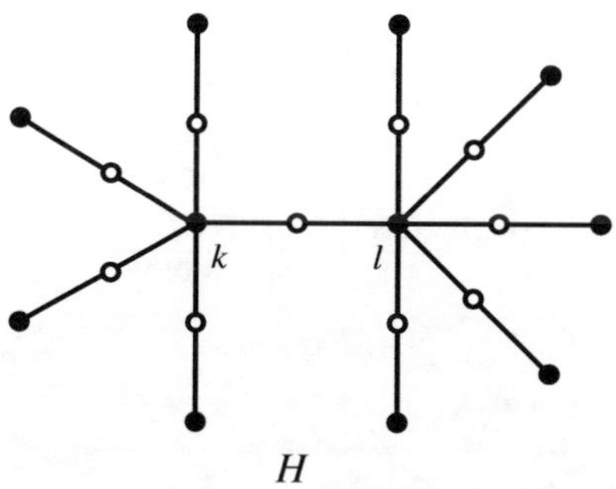

FIGURE 6.9. Series H: ordinary trees of diameter 6 which are decomposable.

6.7.1. Series H. The trees of the series H are compositions of trees from the series C with the parameters $s = t = 1$ and chains of length 2.

The expressions of the Shabat polynomials for the trees from the series C in terms of Jacobi polynomials are given in Section 6.5.5. Using the fact that $s = t = 1$ we can also compute them directly. Indeed, the trees in question have exactly two vertices of degree greater than 1. Putting them into the points $x = 0$ and $x = 1$ and taking into account that the degree of the corresponding Shabat polynomial $S(x)$ is $k + l - 1$,

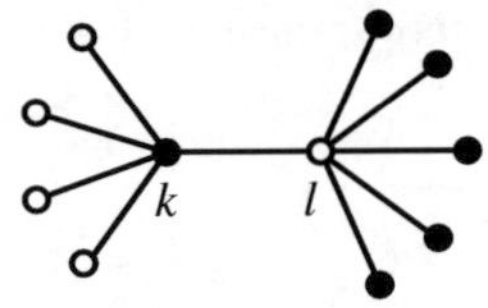

FIGURE 6.10. Replace every edge of this tree with a two-edge chain, and you get the tree H.

we conclude that the derivative of S is proportional to $x^{k-1}(1-x)^{l-1}$. Therefore, the polynomial $S(x)$ itself can be written as

$$S(x) = K \cdot \int_0^x t^{k-1}(1-t)^{l-1}dt. \tag{6.63}$$

Then we automatically have $S(0) = 0$, while in order to get $S(1) = 1$ we must take

$$K = \frac{1}{B(k,l)} = \frac{(k+l-1)!}{(k-1)!\,(l-1)!}, \tag{6.64}$$

where

$$B(k,l) = \int_0^1 t^{k-1}(1-t)^{l-1}dt \tag{6.65}$$

is the Euler beta function.

Then, taking the Shabat polynomial for the chain with two edges and with two black vertices placed at 0 and 1, which is equal to

$$U(y) = 4y(1-y), \tag{6.66}$$

we obtain the following

PROPOSITION 6.18 (Series H). *The polynomial P for the tree H is equal to*

$$P(x) = U(S(x)) \tag{6.67}$$

where U is as in (6.66) *and S is as in* (6.63) *and* (6.64).

The proof is obvious. □

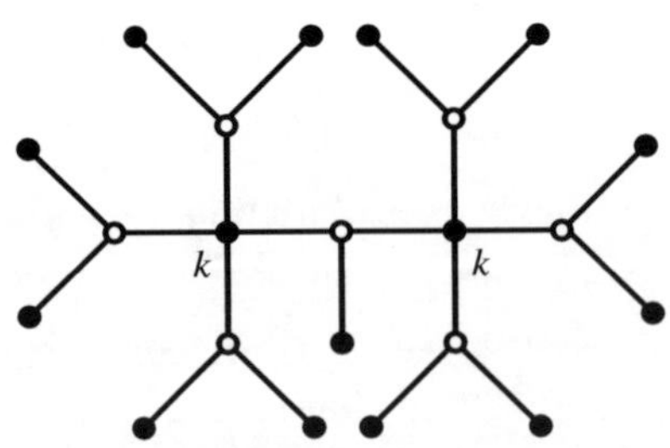

I

FIGURE 6.11. Series I.

6.7.2. Series I. Below are given Shabat polynomials $P(z)$ for the trees of the series I. These trees are compositions of the trees from the series C with $s = t = 1$ and $k = l$, and the stars with three edges. Thus, $P(x) = U(S(x))$, where S is a Shabat polynomial corresponding to a tree from the series C, and U is a Shabat polynomial corresponding to the star with three edges. However, in order to achieve the rationality of the coefficients of P we still must find an appropriate normalization of S.

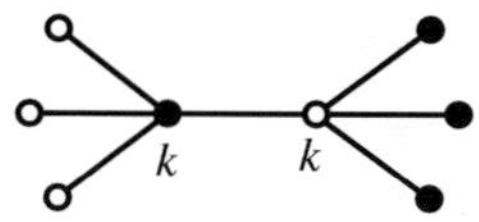

FIGURE 6.12. Replace every edge with a three-edge star, and you get the tree I.

For this purpose, contrary to all traditions, let us put the vertices of degree k of the tree from the series C at the points $x = \pm\sqrt{-3}$. Then the derivative of the corresponding Shabat polynomials $S(x)$ must be equal to

$$S'(x) = a(x+\sqrt{-3})^{k-1}(x-\sqrt{-3})^{k-1} = a(x^2+3)^{k-1}, \quad a \in \mathbb{C}. \tag{6.68}$$

Therefore,

$$S(x) = a\int (x^2+3)^{k-1}dx + b = a\left[\sum_{l=0}^{k-1}\binom{k-1}{l}\frac{x^{2l+1}}{2l+1}3^{k-1-l}\right] + b \tag{6.69}$$

for some $b \in \mathbb{C}$. Substituting the critical points $x = \pm\sqrt{-3}$ into $S(x)$ we obtain the critical values $b \pm c\sqrt{-3}$, where

$$c = a \cdot 3^{k-1}\sum_{l=0}^{k-1}\binom{k-1}{l}\frac{(-1)^l}{2l+1}. \tag{6.70}$$

Setting

$$b = -\frac{1}{2} \tag{6.71}$$

and choosing a in such a way that

$$c = \frac{1}{2}, \tag{6.72}$$

we obtain a polynomial $S \in \mathbb{Q}[x]$ with two critical values

$$y_{1,2} = \frac{-1 \pm \sqrt{-3}}{2}. \tag{6.73}$$

Taking now

$$U(y) = 1 - y^3 \tag{6.74}$$

(we must take $1-y^3$ instead of y^3 in order to get the colors of the vertices which would correspond to Figure 6.11), we obtain the following

PROPOSITION 6.19 (Series I). *The polynomial $P(x)$ for the tree I is equal to*

$$P(x) = U(S(x)), \tag{6.75}$$

where U is as in (6.74) *and S is as in* (6.69) *with a and b defined by conditions* (6.70), (6.71), (6.72).

Once again, the proof is obvious. □

6.8. Series J

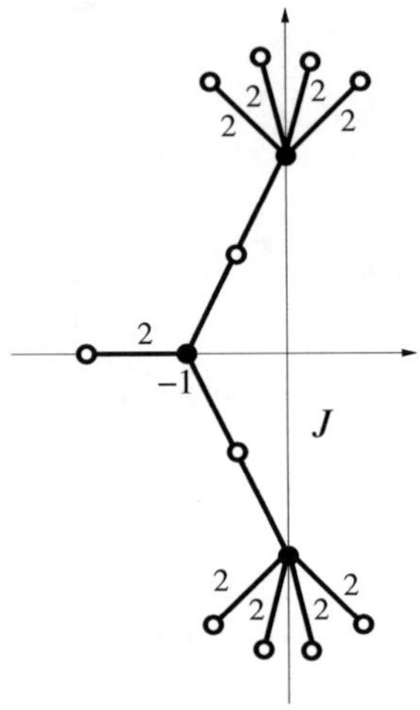

FIGURE 6.13. Series J

This is the last infinite series of unitrees. The number of leaves above and below is the same and is equal to k. Thus, the degree of this tree, or its total weight, is $2k+6$. This tree was shown in Figure 5.7 though now we have changed its position and orientation. In terms of polynomials this means that we normalize the polynomial P as follows:

$$P = (x+1)^4 \cdot (x^2+a)^{2k+1}. \tag{6.76}$$

This means that the black vertex of degree 4 is put at $x = -1$, while two black vertices of degree $2k+1$ are put at the points $\pm\sqrt{-a}$ for certain $a \in \mathbb{Q}$, $a > 0$.

All the white vertices are of degree 2; therefore, the polynomial Q has the form

$$Q(x) = A(x)^2$$

for some polynomial A, $\deg A = 2k+3$. Further, condition (6.2) gives us

$$(x+1)^4 \cdot (x^2+a)^{2k+1} - A(x)^2 \underset{x\to\infty}{=} O(x^{2k+1});$$

here $2k+1$ is the "overweight" of the tree (that is, its total weight minus the number of edges of the topological tree). For the reciprocal polynomials this gives (see (6.1))

$$P^* - Q^* = (1+x)^4 \cdot (1+ax^2)^{2k+1} - A^*(x)^2 \underset{x\to 0}{=} O(x^{2k+5}); \tag{6.77}$$

here $2k+5$ is the number of edges of the topological tree.

PROPOSITION 6.20 (Series J). *The reciprocal polynomials P^* and Q^* for the series J may be represented as follows:*

$$P^* = (1+x)^4 \cdot (1+(2k+4)x^2)^{2k+1}, \quad Q^*(x) = A^*(x)^2, \tag{6.78}$$

where A^* is the initial segment of the series $(P^*)^{1/2}$ up to the degree $2k+3$:

$$(1+x)^2(1+(2k+4)x^2)^{(2k+1)/2} \underset{x\to 0}{=} A^* + O(x^{2k+4}). \tag{6.79}$$

PROOF. Let

$$A^* = (1+x)^2 \cdot (1+ax^2)^{(2k+1)/2} + x^{2k+4} \cdot h \tag{6.80}$$

where

$$h \underset{x\to 0}{=} O(1).$$

Computing $(A^*)^2$ we get

$$\begin{aligned}(A^*)^2 &= P^* + 2x^{2k+4} \cdot h \cdot (1+x)^2 \cdot (1+ax^2)^{(2k+1)/2} + x^{4k+8} \cdot h^2 \\ &= P^* + x^{2k+4}\left[2h \cdot (1+x)^2 \cdot (1+ax^2)^{(2k+1)/2} + x^{2k+4} \cdot h^2\right].\end{aligned} \tag{6.81}$$

Thus, for any value of the parameter a we have

$$P^* - A^*(x)^2 \underset{x\to 0}{=} O(x^{2x+4}),$$

and therefore, in order to obtain (6.77), we only have to show that for $a = 2k+4$ the constant term of h is equal to zero, or, equivalently, the coefficient in front of x^{2k+4} in the series

$$(P^*)^{1/2} = (1+x)^2 \cdot (1+ax^2)^{(2k+1)/2}$$

vanishes.

Let us write the second factor of the latter expression explicitly:

$$\begin{aligned}(1+ax^2)^{(2k+1)/2} &= 1 + \frac{2k+1}{2}ax^2 + \frac{1}{2!}\cdot\frac{(2k+1)(2k-1)}{4}a^2x^4 + \\ &\quad \frac{1}{3!}\cdot\frac{(2k+1)(2k-1)(2k-3)}{8}a^3x^6 + \ldots + \\ &\quad \frac{1}{(k+2)!}\frac{(2k+1)(2k-1)\ldots(-1)}{2^{k+2}}a^{k+2}x^{2k+4} + \ldots\end{aligned} \tag{6.82}$$

Notice that this series involves only even powers. Multiplying it by

$$(1+x)^2 = 1 + 2x + x^2$$

we see that the coefficient in front of x^{2k+4} in $(P^*)^{1/2}$ is the sum of the coefficients in front of x^{2k+4} and x^{2k+2} in (6.82). Therefore, we must ensure that

$$\begin{aligned}&\frac{1}{(k+1)!}\cdot\frac{(2k+1)(2k-1)\ldots\cdot 1}{2^{k+1}}\cdot a^{k+1} + \\ &\frac{1}{(k+2)!}\cdot\frac{(2k+1)(2k-1)\ldots\cdot(-1)}{2^{k+2}}\cdot a^{k+2} = 0.\end{aligned} \tag{6.83}$$

Collecting similar terms we get

$$\frac{1}{(k+1)!}\cdot\frac{(2k+1)(2k-1)\ldots\cdot 1}{2^{k+1}}\cdot a^{k+1}\left(1 + \frac{1}{k+2}\cdot\frac{(-1)}{2}\cdot a\right) = 0, \tag{6.84}$$

which gives $a = 2k+4$. □

EXAMPLE 6.21 (Series J, $k = 3$). Let us take $k = 3$. Then we have:

$$P^* = (1+x)^4(1+10x^2)^7.$$

Further,

$$\begin{aligned}(P^*)^{1/2} &= (1+x)^2(1+10\,x^2)^{7/2}\\ &= 1+2\,x+36\,x^2+70\,x^3+\frac{945}{2}\,x^4+875\,x^5+2625\,x^6+4375\,x^7+\\ &\quad \frac{39\,375}{8}\,x^8+\frac{21\,875}{4}\,x^9-\frac{21\,875}{4}\,x^{11}+\frac{65\,625}{16}\,x^{12}+\ldots\end{aligned}$$

Notice that the term with x^{10} is missing. Finally,

$$\begin{aligned}A^* &= 1+2\,x+36\,x^2+70\,x^3+\frac{945}{2}\,x^4+875\,x^5+2625\,x^6+4375\,x^7+\\ &\quad \frac{39\,375}{8}\,x^8+\frac{21\,875}{4}\,x^9.\end{aligned}$$

6.9. Sporadic trees

As it was explained previously, in Section 6.2, the verification of the results given below is trivial. Therefore, we present nothing else but the polynomials themselves.

6.9.1. Tree K.

$$\begin{aligned}P &= (x^2-5x+1)^3(x^2-13x+49),\\ Q &= (x^4-14x^3+63x^2-70x-7)^2,\\ R &= -1728x.\end{aligned}$$

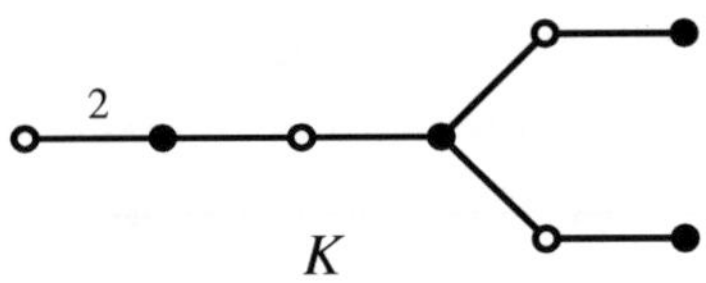

FIGURE 6.14. Tree K.

6.9.2. Tree L.

$$\begin{aligned}P &= (x^3-16x^2+160x-384)^3,\\ Q &= x\,(x^4-24x^3+336x^2-2240x+8064)^2,\\ R &= -2^{14}\cdot 3^3\,(x^2-13x+128).\end{aligned}$$

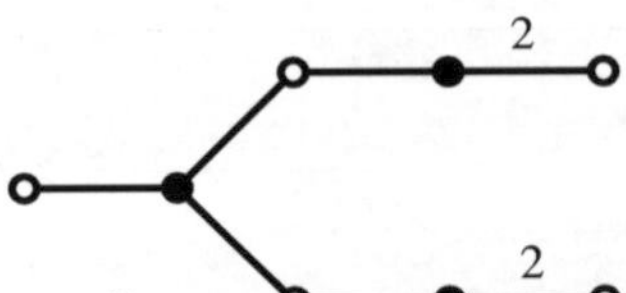

FIGURE 6.15. Tree L.

6.9.3. Tree M.

$$
\begin{aligned}
P &= x(x^3+12x^2+60x+96)^3,\\
Q &= (x^5+18x^4+144x^3+576x^2+1080x+432)^2,\\
R &= -1728\,(3x^2+28x+108).
\end{aligned}
$$

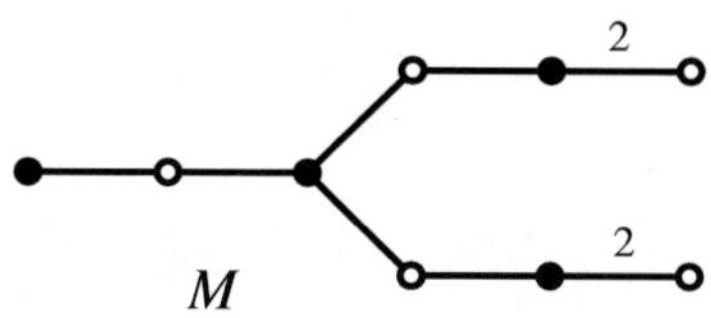

FIGURE 6.16. Tree M.

6.9.4. Tree N.

$$
\begin{aligned}
P &= x^3\,(x^3-8)^3,\\
Q &= (x^6-12x^3+24)^2,\\
R &= 64\,(x^3-9).
\end{aligned}
$$

The tree is symmetric, with the symmetry of order 3. Therefore, P, Q, R are polynomials in x^3.

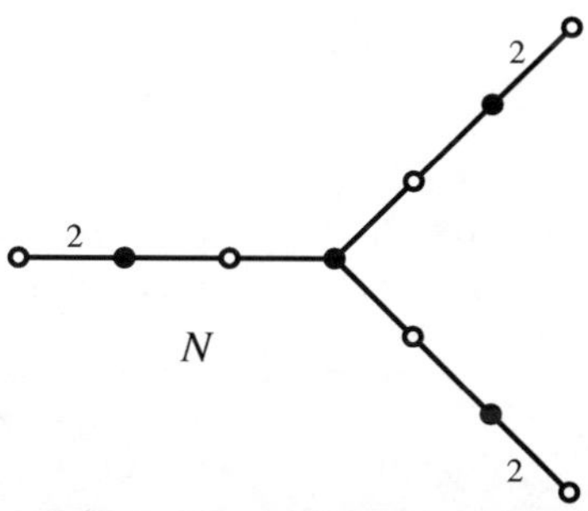

FIGURE 6.17. Tree N.

6.9.5. Tree O.

$$
\begin{aligned}
P &= (x^4+6x^2+64x-55)^5,\\
Q &= (x^{10}+15x^8+160x^7-70x^6+1440x^5+6510x^4\\
&\quad -11\,040x^3+26\,805x^2+40\,160x-226\,797)^2,\\
R &= 2^{20}\,(5x^7+59x^5+690x^4-485x^3+3820x^2\\
&\quad +20\,165x-49\,534).
\end{aligned}
$$

These polynomials were found in Beukers and Stewart [**BeSt-10**]. (In their paper, only the polynomial P is given but it uniquely determines two other polynomials).

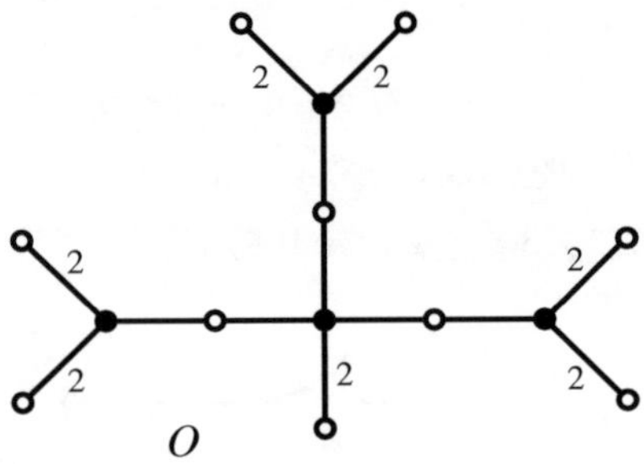

FIGURE 6.18. Tree O.

6.9.6. Tree P.

$$
\begin{aligned}
P &= (x^3 + 9x + 9)^5, \\
Q &= (x^5 + 15x^3 + 15x^2 + 45x + 90)^3, \\
R &= -27\,(15x^8 + 395x^6 + 423x^5 + 3330x^4 + 7290x^3 \\
&\quad + 11\,880x^2 + 29\,565x + 24\,813).
\end{aligned}
$$

Once again, the answer is taken from [**BeSt-10**], with a slight renormalization.

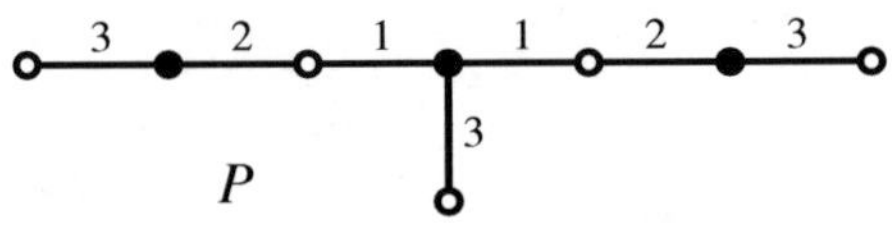

FIGURE 6.19. Tree P.

6.9.7. Tree Q.

This tree is the only sporadic tree from Adrianov's list of ordinary unitrees. Thus, P is a Shabat polynomial, while the polynomial R is a constant.

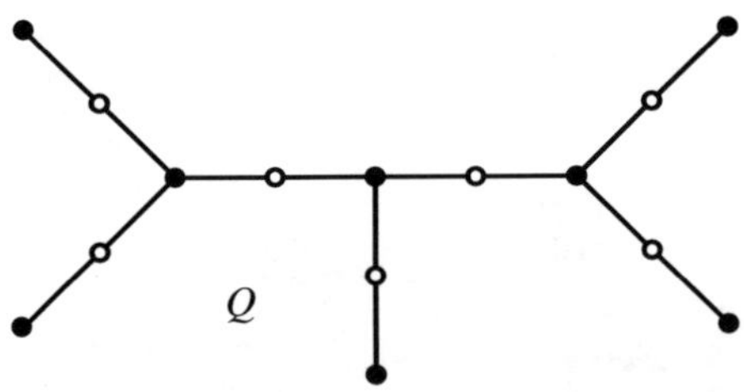

FIGURE 6.20. Tree Q.

$$
\begin{aligned}
P &= (x^3 + 15x + 16)^3(x^5 + 39x^3 + 64x^2 + 384x + 1872), \\
Q &= (x^7 + 42x^5 + 56x^4 + 525x^3 + 1680x^2 + 1792x + 6456)^2, \\
R &= -2^6 \cdot 3^{12}.
\end{aligned}
$$

Note that the positions of certain black vertices are rational:

$$x^3 + 15x + 16 = (x + 1)(x^2 - x + 16),$$
$$x^5 + 39x^3 + 64x^2 + 384x + 1872 = (x + 3)(x^4 - 3x^3 + 48x^2 - 80x + 624).$$

6.9.8. Tree R**.**

The tree R is the "square" of the tree L: it is symmetric, with the symmetry of order 2, and one of its "halves" is equal to L. Therefore, we may take the polynomials for the tree L and insert x^2 instead of x.

$$\begin{aligned} P &= (x^6 - 16x^4 + 160x^2 - 384)^3, \\ Q &= x^2\,(x^8 - 24x^6 + 336x^4 - 2240x^2 + 8064)^2, \\ R &= -2^{14} \cdot 3^3\,(x^4 - 13x^2 + 128). \end{aligned}$$

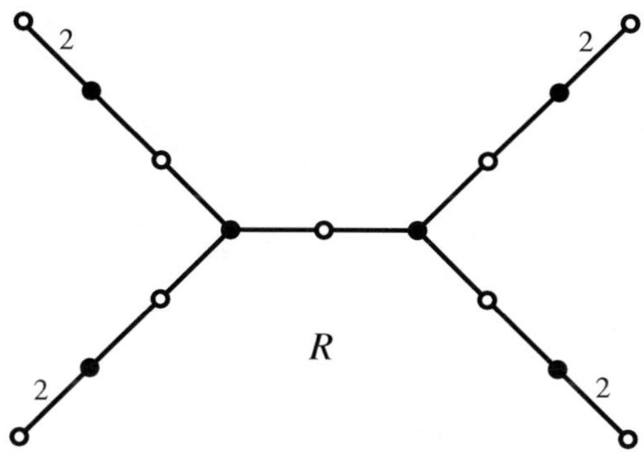

FIGURE 6.21. Tree R.

6.9.9. Tree S**.**

$$\begin{aligned} P &= x^2\,(x^4 + 24x^3 + 176x^2 - 2816)^3, \\ Q &= (x^7 + 36x^6 + 480x^5 + 2304x^4 - 3840x^3, \\ &\quad - 55\,296x^2 - 14\,336x + 221\,184)^2 \\ R &= 2^{22} \cdot 3^3\,(x^3 + 17x^2 + 56x - 432). \end{aligned}$$

Notice that the second factor in P, the one which is "cubed", does not contain the term with x: this is not a misprint.

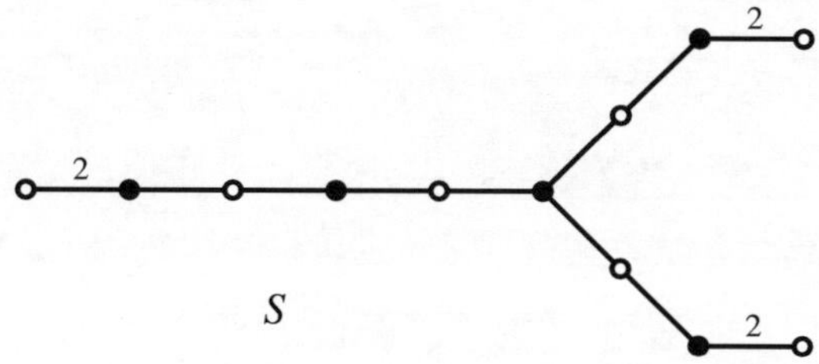

FIGURE 6.22. Tree S.

6.9.10. Tree T.

$$\begin{aligned}
P &= (x^8 + 84x^6 + 176x^5 + 2366x^4 + 13\,536x^3 + 26\,884x^2 \\
&\quad + 218\,864x + 268\,777)^3, \\
Q &= (x^{12} + 126x^{10} + 264x^9 + 6195x^8 + 31\,392x^7 + 163\,956x^6 \\
&\quad + 1\,260\,528x^5 + 3\,531\,639x^4 + 19\,770\,400x^3, \\
&\quad + 62\,912\,622x^2 + 94\,024\,776x + 291\,742\,453)^2, \\
R &= -2^{38} \cdot 3^3\,(x^5 + 62x^3 + 148x^2 + 1001x + 8852).
\end{aligned}$$

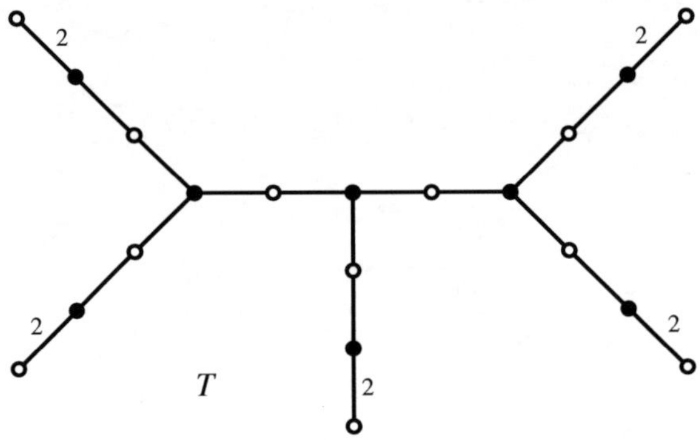

FIGURE 6.23. Tree T.

CHAPTER 7

Primitive monodromy groups of weighted trees

The material of this chapter and of the subsequent one is based on the paper [**AdZv-13**] by Adrianov and Zvonkin.

Let (P, Q) be a DZ-pair of polynomials, and let $R = P - Q$. The Belyĭ function $f = P/R$, $f : \overline{\mathbb{C}} \to \overline{\mathbb{C}}$, may be considered as a ramified covering of the Riemann sphere $\overline{\mathbb{C}}$ by itself. The *monodromy group* of this covering is a powerful and important invariant of the Galois action on dessins and, in particular, on weighted trees. The monodromy group corresponding to a dessin is sometimes called *cartographic group*, or else *edge rotation group*; the origin of the latter term will become clear below.

In this chapter, and in the subsequent one, we classify all weighted trees with *primitive* monodromy groups. Apart from the four trivial cases S_n, A_n, C_p and D_p (with p prime), there are, up to the color exchange, 184 such trees, which are subdivided into (at least) 85 Galois orbits and which generate 34 primitive groups (the highest degree of a group in this list is 32). This result may also be considered as a modest contribution to the classification of coverings of genus zero with primitive monodromy groups.

In our book, in view of its size, we cannot explain the entire motivation behind the primitive groups. We only point out that primitive groups are few: see Remark 7.6, Theorem 7.7 and Table 1 (pages 89–90).

7.1. Monodromy group of a dessin

DEFINITION 7.1 (Monodromy group of a dessin). Label the n edges of a bicolored map by numbers from 1 to n respecting Convention 7.2 below, and write down two permutations a and b as follows. The cycles of a indicate the cyclic order of edges, in the counter-clockwise direction, around the black vertices, and the cycles of b indicate the cyclic order of edges, in the counter-clockwise direction, around the white vertices. Then the *monodromy group* (also called the *edge rotation group* or the *cartographic group*) of the map in question is the permutation group $G = \langle a, b\rangle$ of degree n. This group is defined up to a conjugacy inside S_n.

Note the following properties of the above group.

- The action of the monodromy group on the edges of a map is transitive since by definition a map is connected.
- The cycle structures of a and b are, respectively, α and β, the partitions corresponding to the degrees of black and white vertices.
- An important property: the faces of the map are determined by the cycles of the permutation $c = (ab)^{-1}$ (so that $abc = 1$).
- The monodromy group of a dessin is isomorphic to the monodromy group of the ramified covering of the sphere realized by the Belyĭ function corresponding to the map in question.

CONVENTION 7.2 (Placing the labels). For a given edge, it is convenient to put its label on the left bank of this edge while moving from the black end of the edge to the white one. In this way, the labels of the cycles of the permutation c describing the faces will always be situated inside the corresponding faces. Moreover, they will go around the face in the counter-clockwise direction when we look at them from the center of the face.

(For the outer face of a plane map, its center is situated "behind the sphere". Therefore, an observer looking at the map from a position "in front of the sphere" will have an impression that the labels go around the outer face in the clockwise direction. This is an illusion: *looking from the center of the outer face*, the labels go counter-clockwise.)

All the above properties are valid for arbitrary bicolored maps on oriented surfaces, not necessarily plane and not necessarily corresponding to the weighted trees.

EXAMPLE 7.3 (Labelling of a map). Figure 7.1 shows the same map as in Figure 1.1 (page 5), with its 18 edges labelled by numbers from 1 to 18. The corresponding permutations are as follows (for the readers' convenience, we also indicate the fixed points of these permutations):

$$\begin{aligned} a &= (1)(2,9,12,13,3)(4,6)(7)(5,18,17,16,15)(8,14)(10,11), \\ b &= (1,10,11,9)(2)(3,13,12,8,14,7,6)(4,5,15,16,17,18), \\ c &= (1,2,3,4,5,6,7,8,9,10)(11)(12)(13)(14)(15)(16)(17)(18). \end{aligned}$$

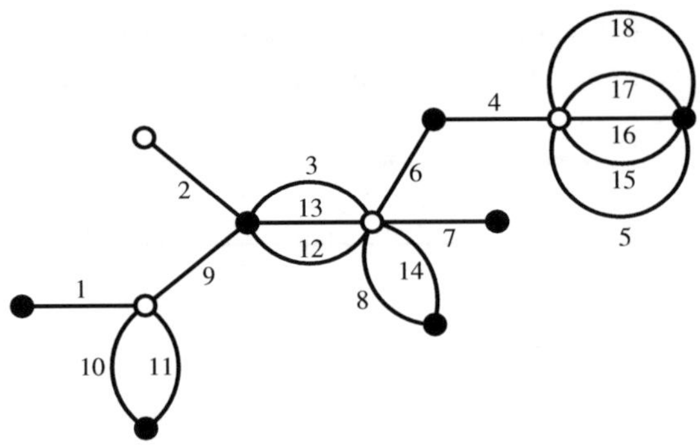

FIGURE 7.1. A bicolored map with labeled edges.

Having represented a bicolored map as a triple of permutations (a, b, c) such that $abc = 1$ we may introduce two *braid operations*:

$$\sigma_1 : (a,b,c) \to (b, b^{-1}ab, c), \qquad \sigma_2 : (a,b,c) \to (a, c, c^{-1}bc).$$

Notice that these operations preserve the product of the three permutations in a triple. They also satisfy the braid relation $\sigma_1\sigma_2\sigma_1 = \sigma_2\sigma_1\sigma_2$. Obviously, all the triples obtained by braiding generate the same group.

It is natural to call the operation σ_1 *color exchange* since now the permutation b becomes the first element of the triple and therefore corresponds to the black vertices, the permutation $b^{-1}ab$ conjugate to a corresponds to the white vertices, and the permutation c corresponding to faces remains the same.

The operation σ_2 exchanges white vertices with faces (the face permutation becomes conjugate to b) and preserves the black vertices. The operation $\sigma_1\sigma_2\sigma_1$ gives the triple $(c, c^{-1}bc, c^{-1}b^{-1}abc) = (c, c^{-1}bc, a)$ since $abc = 1$. Thus, the operation $\sigma_1\sigma_2\sigma_1$ exchanges black vertices with faces while preserving the white vertices (the white permutation becomes conjugate to b). It is natural to call both operations σ_2 and $\sigma_1\sigma_2\sigma_1$ *dualities*. A map which is isomorphic to one of its duals is called *self-dual*.

7.2. Primitive and imprimitive groups

DEFINITION 7.4 (Imprimitive, primitive and special groups). A transitive permutation group G of degree n acting on a set X, $|X| = n$, is called *imprimitive* if the set X can be subdivided into m disjoint *blocks* $X_1, \ldots, X_m$ of equal size $|X_i| = n/m$, where $1 < m < n$, such that an image of a block under the action of any element of G is once again a block. A transitive permutation group which is not imprimitive is called *primitive*. We will call a permutation group *special* if it is primitive and not equal to S_n or A_n.

THEOREM 7.5 (Ritt's theorem). *A ramified covering is a composition of two or more coverings of smaller degrees if and only if its monodromy group is imprimitive.*

Ritt proved only a particular case of this theorem. The reader may look for a proof in Section 1.7.2 of the book [**LaZv-04**]; for a better proof see ERRATA AND COMMENTS to this book.

REMARK 7.6 (Primitive groups are few). There are 301 transitive permutation groups of degree $n = 12$, but only six of them are primitive (S_n and A_n included); there are (exactly) 25 000 transitive permutation groups of degree $n = 24$, but only five of them are primitive. The information about the numbers of primitive and transitive groups may be found in [**Hul-05**] (up to $n = 32$) and in the database [**Gal-DB**] (up to $n = 47$ but except $n = 32$). We sum up this information in Table 1 below.

When n is prime, all groups of degree n are primitive, but they are not numerous as well. To compare, there are 1854 transitive groups of degree 28, only 8 groups of degree 29, and 5712 groups of degree 30.

In fact, for the majority of $n \in \mathbb{N}$ there are only two primitive permutation groups of degree n, namely, S_n and A_n. The following theorem belongs to Cameron, Neumann and Teague [**CNT-82**][1]:

THEOREM 7.7 (Only S_n and A_n). *Let $t(N)$ denote the number of integers $n < N$ for which there is a primitive group of degree n other than S_n or A_n. Then*

$$\frac{t(N)}{N} \sim \frac{2}{\ln N} \to 0 \quad \text{as} \quad N \to \infty.$$

Notice that this theorem counts degrees, not groups. To give but one example, for $n = p$ prime there exist, beside other possible groups, the collection of groups of the type $\mathrm{C}_p \rtimes \mathrm{C}_k$ for all k divisors of $p - 1$. Here $\rtimes$ stays for the semidirect product; for $k = 1$ we get the group C_p itself, and for $k = p - 1$ we get the affine group $\mathrm{AGL}_1(p)$. Nevertheless, this n is counted only once.

[1]We are grateful to Gareth Jones who indicated to us this reference.

TABLE 1. The number of primitive and transitive groups of degree $n \leq 47$. The data is taken from the [**Hul-05**] and [**Gal-DB**].

Degree	Primitive	Transitive	—	Degree	Primitive	Transitive
2	1	1		25	28	211
3	2	2		26	7	96
4	2	5		27	15	2392
5	5	5		28	14	1854
6	4	16		29	8	8
7	7	7		30	4	5712
8	7	50		31	12	12
9	11	34		32	7	2 801 324
10	9	45		33	4	162
11	8	8		34	2	115
12	6	301		35	6	407
13	9	9		36	22	121 279
14	4	63		37	11	11
15	6	104		38	4	76
16	22	1954		39	2	306
17	10	10		40	8	315 842
18	4	983		41	10	10
19	8	8		42	4	9491
20	4	1117		43	10	10
21	9	164		44	4	2113
22	4	59		45	9	10 923
23	7	7		46	2	56
24	5	25 000		47	6	6

Thus, permutation groups are mostly imprimitive. But this does not mean that we encounter them more often. Indeed, a group generated by a pair of randomly chosen permutations is either S_n or A_n with a probability which tends to 1 as $n \to \infty$ (see [**Bab-89**]). Roughly speaking, we may say that first come S_n and A_n; then, rarely, imprimitive groups; and then, exceptionally, special groups.

A well-known conjecture by Guralnick and Thompson [**GuTh-90**] states that for any g the *composition factors* of monodromy groups of genus g coverings of the sphere are either alternating groups A_n, $n \geq 5$, or cyclic groups C_p for prime p, or that they belong to a finite list of exceptional groups. The conjecture was proved by Frohardt and Magaard in [**FrMa-01**].

Nevertheless the finiteness statement for simple groups which may appear as composition factors does not guarantee a similar statement about possible primitive monodromy groups of coverings.[2] There is a stronger and more subtle conjecture by Guralnick [**GuSh-07**] concerning primitive monodromy groups. It is known that there are families of special primitive monodromy groups for $n = d(d-1)/2$ or

[2]Unfortunately, the meaning of this conjecture was misinterpreted in the paper [**AdZv-13**] and in the Russian edition of the book [**LaZv-04**].

$n = d^2$ (in these cases the composition factors are still either cyclic or alternating A_d, but we consider them as special since $d < n$).

To the best of our knowledge, the latter conjecture is not proved and a complete classification of the primitive monodromy groups of genus 0 coverings of the sphere is not yet achieved. There are, however, some partial results, like a complete list of the affine groups appearing in this context which is given in [**MSW-11**], or the upper bound $n \leq 10\,000$ for the degree of classical groups as monodromy groups of genus $g \leq 2$ which is obtained in [**FGM-14**]. Our result may also be considered as a small step in the same direction.

We already know that the passport of a dessin is an invariant of the Galois action. The monodromy group is another important invariant. Thus, if in the same combinatorial orbit there are dessins with different monodromy groups then this combinatorial orbit splits into several Galois orbits. Even more so, the character table of the group contains an important information about the field of moduli of the dessin. These considerations give us an additional impetus to study monodromy groups.

7.3. A menagerie of groups

For the reader's convenience, we list here the standard notation for permutation groups appearing in our exposition; some of them were already used before.

COMMON GROUPS

- S_n is the *symmetric group* acting on n points, that is, the group containing all permutations of n points.
- A_n is the *alternating group* acting on n points, that is, the group containing the even permutations of n points.
- C_n is the *cyclic group* of order n.
- D_n is the *dihedral group* of order $2n$, which is the group of all the symmetries of a regular polygon with n sides.

LINEAR GROUPS

Below, p denotes a prime, $q = p^e$, $e \geq 1$, is a prime power, $\mathbb{F}_q$ is the finite field with q elements, and d is a dimension. The automorphism group of the field $\mathbb{F}_q$ is the cyclic group C_e. This group is generated by the *automorphism of Frobenius* $a \mapsto a^p$. (It is an automorphism since $(ab)^p = a^p b^p$, and the binomial formula implies that $(a+b)^p = a^p + b^p$.)

- $\mathrm{GL}_d(q)$ is the *general linear group*, that is, the group of non-degenerate linear transformations of the d-dimensional vector space $\mathbb{F}_q^d$ over $\mathbb{F}_q$, or, equivalently, the group of non-degenerate $d \times d$ matrices with the elements in $\mathbb{F}_q$.
- $\mathrm{SL}_d(q)$ is the *special linear group*, that is, the subgroup of $\mathrm{GL}_d(q)$ containing the matrices of the determinant 1.
- $\Gamma\mathrm{L}_d(q)$ is the *semilinear group*; its elements are called *semilinear transformations*, or *collineations*; they are the mappings $f : \mathbb{F}_q^d \to \mathbb{F}_q^d$ satisfying the conditions

$$f(x+y) = f(x) + f(y), \quad f(\lambda x) = \lambda^\sigma x,$$

where σ is a power of the automorphism of Frobenius.

PROJECTIVE GROUPS

- $\mathrm{PGL}_d(q)$ is the *projective general linear group*, that is, the quotient of $\mathrm{GL}_d(q)$ by the subgroup of scalar matrices $\lambda \cdot \mathrm{Id}$, $\lambda \in \mathbb{F}_q^*$, $\mathrm{Id} \in \mathrm{GL}_d(q)$.
- $\mathrm{PSL}_d(q) = \mathrm{L}_d(q)$ is the *projective special linear group*, that is, the group containing the elements of $\mathrm{PGL}_d(q)$ having a representative with the determinant equal to 1.
- $\mathrm{P\Gamma L}_d(q)$ is the *projective semilinear group*, that is, the quotient of the group $\Gamma\mathrm{L}_d(q)$ by the subgroup of scalar matrices.

AFFINE GROUPS

- $\mathrm{AGL}_d(q)$ is the *affine general linear group*, that is, the group of the affine transformations of the space $\mathbb{F}_q^d$:

$$\{x \mapsto Ax + b \mid A \in \mathrm{GL}_d(q),\ b \in \mathbb{F}_q^d\}.$$

- $\mathrm{ASL}_d(q)$ is the *affine special linear group*, that is, the subgroup of $\mathrm{AGL}_d(q)$ in which $A \in \mathrm{SL}_d(q)$.
- $\mathrm{A\Gamma L}_d(q)$ is the *affine semilinear group*, which means that $A \in \Gamma\mathrm{L}_d(q)$.

MATHIEU GROUPS

- $\mathrm{M}_{11}, \mathrm{M}_{12}, \mathrm{M}_{22}, \mathrm{M}_{23}$ and M_{24} are *Mathieu groups*. They are the first five groups of the famous list of 26 sporadic finite simple groups. A detailed definition of the Mathieu groups would lead us too far away. An interested reader should consult a group-theoretic literature.

The three projective groups $\mathrm{PSL}_d(q)$, $\mathrm{PGL}_d(q)$ and $\mathrm{P\Gamma L}_d(q)$ act on the projective space of dimension $d-1$ over the field $\mathbb{F}_q$. The group $\mathrm{P\Gamma L}_d(q)$ is the full automorphism group of this space. The number of points in this space, and thus the degree of all the three groups, is equal to $(q^d-1)/(q-1) = 1+q+q^2+\ldots+q^{d-1}$. These three groups form a chain of normal subgroups

$$\mathrm{PSL}_d(q) \trianglelefteq \mathrm{PGL}_d(q) \trianglelefteq \mathrm{P\Gamma L}_d(q).$$

The quotient of $\mathrm{P\Gamma L}_d(q)$ by $\mathrm{PGL}_d(q)$ is isomorphic to C_e. Therefore, if $q = p$ is prime then the two groups coincide. The quotient of $\mathrm{PGL}_d(q)$ by $\mathrm{PSL}_d(q)$ is isomorphic to the quotient $\mathbb{F}_q^*/\{a^d \mid a \in \mathbb{F}_q^*\}$ where $\mathbb{F}_q^* = \mathbb{F}_q \setminus \{0\}$ is the cyclic multiplicative group of non-zero elements of $\mathbb{F}_q$. The size of this quotient depends on the numeric relations between q and d. In certain cases the groups $\mathrm{PSL}_d(q)$ and $\mathrm{PGL}_d(q)$ may coincide as well.

The notation $G : H$ (see orbit 17.2 on page 126) means *split extension* of G by H, and the notation $G.H$ (see orbits from 16.1 to 16.4 on pages 123–125) means *non-split extension* of G by H. Once again, to learn the meaning of these notions we recommend the reader to consult the group-theoretic literature.

7.4. Primitive groups containing a cycle: Gareth Jones's classification

The monodromy groups of weighted trees, that is, of plane maps with all their faces except the outer one being of degree 1, contain permutations with the cycle structure $(n-t)^1 1^t$. Motivated by our study of weighted trees, Gareth Jones classified all special permutation groups containing such a permutation, see [**Jon-14**]. In particular, he showed that in all such cases the number of fixed points is $t \leq 2$. This property is based on two results. The first is an old theorem by Jordan [**Jor-1871**] stating that a primitive group containing a permutation with the cycle structure $(n-t)^1 1^t$ is $(t+1)$-transitive. The second is the complete classification of multiply transitive groups: it is based on the *supertheorem* of the classification of finite simple groups.

We would be happy to present a proof of Jones's result, especially since his paper is not long. Unfortunately, there is no way to do that in a self-contained way. The classification of finite groups looks like a kind of a mushroom spawn which can go on for kilometers while a particular theorem resembles a single mushroom. Thus, in his apparently short paper, Jones uses his own previous results, and also those by Feit, and by Bubboloni and Praeger, and by Müller (a paper of some 80 pages), and so on. All this is, of course, with the *supertheorem* taken for granted. We would need a separate book in order to give a full account of all these results.

The classification due to Jones looks as follows:

THEOREM 7.8 (G. A. Jones [**Jon-14**]). *Let G be a primitive permutation group of degree n, not equal to S_n or A_n. Suppose that G contains a permutation with cycle structure $(n-t)^1 1^t$. Then $t \le 2$, and one of the following holds:*

0. *$t = 0$ and either*
   (a) *$C_p \le G \le AGL_1(p)$ with $n = p$ prime, or*
   (b) *$PGL_d(q) \le G \le P\Gamma L_d(q)$ with $n = (q^d - 1)/(q-1)$ and $d \ge 2$ for some prime power $q = p^e$, or*
   (c) *$G = L_2(11)$, M_{11} or M_{23} with $n = 11$, 11 or 23 respectively.*
1. *$t = 1$ and either*
   (a) *$AGL_d(q) \le G \le A\Gamma L_d(q)$ with $n = q^d$ and $d \ge 1$ for some prime power $q = p^e$, or*
   (b) *$G = L_2(p)$ or $PGL_2(p)$ with $n = p+1$ for some prime $p \ge 5$, or*
   (c) *$G = M_{11}$, M_{12} or M_{24} with $n = 12$, 12 or 24 respectively.*
2. *$t = 2$ and $PGL_2(q) \le G \le P\Gamma L_2(q)$ with $n = q+1$ for some prime power $q = p^e$.*

The set of groups listed in the above theorem is infinite. When, in addition to the existence of elements with cycle structure $(n-t)^1 1^t$, we impose a planarity condition, that is, if we want to get such a group as the monodromy group of a plane map with the face degrees $(n-t)^1 1^t$, then we get a finite set of groups, with two exceptions considered below, namely, C_p and D_p with p prime.

7.4.1. Cyclic and dihedral groups. For $n = p$ prime, all transitive groups of degree p are primitive. According to Theorem 7.8, all the groups G between the cyclic group C_p and the affine group $AGL_1(p)$ contain a cycle of length p and therefore, in principle, might be monodromy groups of ordinary trees. But, according to [**AKS-97**], only two of them are indeed realized by trees, see Figure 7.2:

- the cyclic group C_p of order p, which is the monodromy group of the star-tree with p edges;
- the dihedral group D_p of order $2p$, which is the monodromy group of the chain-tree with p edges.

In what follows, we avoid mentioning these two cases.

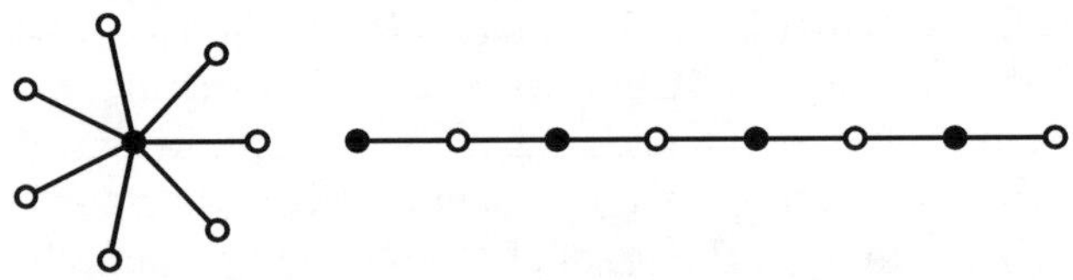

FIGURE 7.2. The monodromy group for the star-tree is the cyclic group C_p, and for the chain-tree it is the dihedral group D_p. If p is prime both groups are primitive.

Thus, among the subgroups of $AGL_1(p)$ only the group $AGL_1(p)$ itself remains of interest for us since, according to the case 1(a) of Theorem 7.8, this group contains permutations with the cycle structure $(p-1)^1 1^1$, and therefore it could be the monodromy group of a map with p edges, one face of degree 1, and p vertices. However, it is easy to see that this can only happen for $p = 5$ and $p = 7$. Indeed, the cycle structures of the elements of $AGL_1(p)$ are all of the form $l^k 1^1$ where $lk = p-1$. This partition has $k+1$ parts, and a pair of such partitions has $k_1 + k_2 + 2$ parts, $k_1, k_2 \geq 1$ being divisors of $p-1$. Now, if $k_1 = k_2 = (p-1)/2$, both partitions become equal to $2^k 1^1$, and we get a chain tree. Otherwise, the biggest possible number of parts in two partitions (which should give us the number of vertices) is $m = \dfrac{p-1}{2} + \dfrac{p-1}{3} + 2$, and the inequality $m \geq p$ leads to $p \leq 7$.

7.5. Main theorem: statement and commentaries

THEOREM 7.9 (Classification of special weighted trees). *The complete list of special weighted trees, not taking into account stars and chains, is given in the tables of* Chapter 8. *It contains, up to the color exchange,* 184 *trees, which are subdivided into (at least)* 85 *Galois orbits and generate* 34 *different monodromy groups.*

REMARK 7.10 (Color exchange). One should be careful when counting trees "up to the color exchange" in the case when $\alpha = \beta$, that is, when the black partition is equal to the white one. There are several such orbits in our list: 6.7, 8.1, 8.7, 8.9, 9.7, 12.10, 12.12, 24.5 (see the group generators and the pictures in the chapter to follow). Almost all of them are invariant under the color exchange, except the trees in orbits 8.1 (see Figure 8.6, page 104) and 9.7 (see Figure 8.13, page 111). In 8.1 the color exchange gives a different tree in the same orbit. In 9.7 the combinatorial orbit contains four trees; two of them are invariant under the color exchange while the other two are not.

REMARK 7.11 (Duality). There is a number of dual and self-dual weighted trees in our list:

- dessins 5.1, 6.2, 6.9, 8.5, 8.14 and 10.3 are self-dual;
- dessins 6.3 and 6.4, 6.5 and 6.6, 6.7 and 6.8, 8.12 and 8.13, two dessins in orbit 9.5 are pairwise dual.

REMARK 7.12 (Galois orbits). The complete set of combinatorial Galois invariants most probably does not exist (see in this respect the discussion in Chapter 10). Observe, for example, the following two cases: 9.5 and 9.7. The combinatorial orbit 9.7 consists of 4 trees. The monodromy group of all of them is the same. However, as it was told above, two trees are invariant under the color exchange while the other two are not. This gives us a combinatorial explanation of the splitting of this combinatorial orbit into two Galois orbits. At the same time we do not see any combinatorial (or any other) reason to explain the splitting of the would-be "orbit" 9.5. The monodromy group detaches two trees out of six in the combinatorial orbit, but these two trees still split into two Galois orbits both defined over $\mathbb{Q}$.

We have computed all fields of moduli for $n \leq 12$ except for the orbits 12.3 and 12.9. Also, Belyĭ functions for certain ordinary trees with $n > 12$ edges were computed in [**Mat-96**] (for $n = 23$) and [**CaCo-99**] (for $n = 21, 31$). Yet more fields were computed at our demand by Yuri Matiyasevich, John Voight and Alexander Vatuzov (personal communications), to whom we express our gratitude. There remain, however, several cases for $n > 12$ when we cannot guarantee that, what appears as an "orbit" in our list, is indeed a single Galois orbit and not a union of several Galois orbits. That is why we say with caution "*at least* 85 Galois orbits".

Sometimes we still can say something about fields of moduli without explicit computations:

- When a dessin is invariant under the color exchange then we can express its Belyĭ function via a Belyĭ function of another dessin. For example, let f_1 be the Belyĭ function for dessin 6.7 and f_2, for dessin 6.5. Then $f_1(z^2) = (2f_2(z) - 1)^2$. In particular their fields of moduli are equal. The same is true for 8.2 and 8.7, 8.8 and 8.9, 9.6 and 9.7(a), 12.5 and 12.10, 12.6 and 12.12, 24.1 and 24.5.
- Chapter 8 is a very good illustration of the intimate relations between character tables and fields of moduli. Sometimes the information contained in a character table permits to gain a new knowledge. For example, we could never guess that the polynomial of degree 10 in the orbit 24.4 factorizes over the field $\mathbb{Q}(\sqrt{-23})$ if this field was not present in the character table of the group M_{24}.

7.6. Main theorem: proof

The case $t = 0$, that of *ordinary trees*, was already settled in [**AKS-97**]; the complete list of the corresponding 48 trees may be found there (we also list them in our tables).

All genus zero generating sets for the *affine groups* (case 1(a) of Theorem 7.8) are listed in [**MSW-11**]; see also [**Wan-11**]. Taking the generating sets containing the *triples* of permutations, one of which has a cycle structure $(n-1)^1 1^1$, permits us to settle this case.

The case 1(c) of Theorem 7.8, that is, the case of Mathieu groups M_{11}, M_{12} and M_{24}, can easily be handled using the GAP computer system.

All the remaining groups (the cases 1(b) and 2 of Theorem 7.8) are subgroups of $P\Gamma L_2(q)$ for some prime power $q = p^e$, acting on $n = q + 1$ points. Notice the subscript 2: the projective geometry in question is always a projective *line*.

We will need the following lemma which was proved in [**Mul-95**] and, independently, in [**Adr-97b**]. We state it without proof.

LEMMA 7.13 (Number of cycles). (1) *Let $q = p^e$ be a prime power, $e > 1$, and let $h > 1$ be the least prime divisor of e. Then a non-identity permutation $g \in \mathrm{P\Gamma L}_2(q)$ cannot have more than $l = p^{e/h} + 1$ fixed points.*

(2) *A non-identity permutation $g \in \mathrm{P\Gamma L}_2(p)$ with p prime cannot have more than two fixed points.*

(3) *If no element of a permutation group of degree n has more than l fixed points then an element of order k can have at most $\dfrac{n-l}{k} + l$ cycles.* □

A map with n edges and at most two faces of degree 1 must have at least $n - 1 = q$ vertices. If both permutations a and b are involutions we get a chain-tree. If at least one of them is of order $k \geq 3$ and thus have, roughly, n/k cycles, then, once again roughly, we get no more than $5n/6$ vertices, which is not enough.

More exactly, for $q = p^e$, $e > 1$, denote $u = \sqrt{q}$, so that $q = u^2$. According to the above lemma, a non-identity element of $\mathrm{P\Gamma L}_2(q)$ cannot have more than $u + 1$ fixed points, and thus, according to the third statement of the above lemma, an element of order k cannot have more than

$$\frac{(q+1)-(u+1)}{k} + (u+1)$$

cycles. Therefore, the total number of cycles in two permutations is not greater than

$$\frac{q-u}{2} + (u+1) + \frac{q-u}{3} + (u+1) = \frac{5}{6}(u^2 - u) + 2(u+1).$$

The quadratic inequality

$$\frac{5}{6}(u^2 - u) + 2(u+1) \geq u^2$$

leads to

$$u \leq \frac{7+\sqrt{97}}{2} < \frac{17}{2} = 8.5$$

so that $q \leq 8.5^2 = 72.25$. The biggest prime power satisfying this inequality is 64.

In the same way, for the group $\mathrm{P\Gamma L}_2(p)$ with p prime we have

$$\frac{(p+1)-2}{2} + 2 + \frac{(p+1)-2}{3} + 2 = \frac{5(p-1)}{6} + 4 \geq p,$$

which leads to $p \leq 19$.

There remains a finite number of groups to study. Some of them can be easily ruled out by hand, see Example 7.14 below. The remaining cases were treated by GAP. Theorem 7.9 is proved. □

EXAMPLE 7.14 (Groups eliminated "by hand"). Let us take $q = 64 = 2^6$. According to Lemma 7.13, we have $h = 2$, so that the upper bound for the number of fixed points is $2^{6/2}+1 = 9$. The number of cycles in an involution is then bounded by $(64-8)/2+8 = 36$, and the number of cycles in a permutation of order 3 is bounded by $(64-7)/3+7 = 26$; for all the other permutations in this group the number of cycles is even less than that. However, $36 + 26 = 62$ vertices are not enough in order to create a tree with $n = q + 1 = 65$ edges and with one or two faces of degree 1.

Similar considerations permit us to handle the case $q = 27 = 3^3$. Here $h = 3$, so that the number of fixed points is bounded by $3^{3/3} + 1 = 4$, and the total number of cycles in two permutations, at least one of which is not an involution, is bounded by

$$(27-3)/2 + 3 + (27-3)/3 + 3 = 26,$$

which is not enough to create a tree with 28 edges and with one or two faces of degree 1.

Note, however, that this simple approach fails for $q = 49$. The number of fixed points is bounded by $7^{2/2} + 1 = 8$; the total number of cycles in two permutations is bounded by

$$(49 - 7)/2 + 7 + (49 - 7)/3 + 7 = 49.$$

This number of vertices is *a priori* sufficient in order to create a tree with 50 edges and with two faces of degree 1. Therefore, in order to rule out the group $\mathrm{P\Gamma L}_2(49)$ we need to study it using GAP.

REMARK 7.15 (Cross-verification). As in any experimental work, in order to be on the safe side it is useful to attack the problem from various perspectives and to see if the results thus obtained are coherent. In our work, we did the following:

- We looked through *all* primitive groups up to degree 127 (not only those listed in Theorem 7.8) using the GAP system. Thus, Jones's classification has also received an independent experimental confirmation.
- Many, though not all of the above groups were also studied using the Maple package `group`.
- For the groups of degree up to 11 we used the catalogue [**BMK-83**], and for the groups of degree 14 and 15, the catalogue [**But-93**].
- Whenever possible, we have computed Belyĭ functions in order to verify if indeed a combinatorial orbit in question splits into several Galois orbits, and if one of the orbits does correspond to a group from our list. For example, there exist 16 trees of weight 10 with the passport $(8^1 1^2, 2^4 1^2, 8^1 1^2)$. They split into *four* Galois orbits (see Example 9.23, page 151), and one of them does consist of a single tree corresponding to the group $\mathrm{PGL}_2(9)$: see the orbit 10.3.
- For a given passport, the number of distinct solutions given by Maple was compared with the total number of trees in the corresponding combinatorial orbit. The latter was either constructed using GAP, or computed using the character tables of S_n.

CHAPTER 8

Trees with primitive monodromy groups

The GAP system contains a data library of all primitive permutation groups of degree up to 2499. The command PrimitiveGroup(n,k) returns the kth element of the list of primitive groups of degree n.

When there is a pair of trees mirror symmetric to each other we present only one of them, both in a permutation form and as a figure. Such trees necessarily belong to the same Galois orbit since they can be obtained from each other by a complex conjugation, which is an element of the absolute Galois group $\Gamma = \mathrm{Aut}(\overline{\mathbb{Q}}|\mathbb{Q})$.

For every degree, the groups are sorted according to their position in the GAP library of primitive groups (the developers guarantee that the positions of the groups will not be changed in the future). The orbits are numbered by a double index of the form $n.i$ where n is the degree and i is the number of the orbit among those of degree n. For example, an orbit number 6.8 is the 8th orbit in the list of the orbits of degree 6. The entries like PrimitiveGroup(7,5) are GAP commands to access to the corresponding group (in this case, to the group $L_3(2) = PSL_3(2)$).

The permutation a always corresponds to the black vertices, and b corresponds to the white ones. Every figure contains a distinguished edge: it is the *root edge*, which means that it is labeled by 1. This permits to easily establish the correspondence between permutations and figures. The choice of the root edge does not always look geometrically natural but it is not our choice: the permutations were found by some random search using GAP.

Fields of moduli are computed for all the trees of degree $n \le 11$, for the majority of them of degree 12, and for certain trees of greater degrees. Some fields are computed by us; others, by our colleagues at our request. We also use the information found in publications of various authors. Nevertheless, the information we provide remains incomplete (though always quite plausibly conjectured).

For every group, we provide an information about algebraic numbers present in their character tables (if there are any). It is well known that all the corresponding fields are cyclotomic, that is, generated by roots of unity.

NOTATION 8.1 (Roots of unity). The first nth primitive root of unity will be denoted by e_n:

$$e_n = \cos(2\pi/n) + i\sin(2\pi/n).$$

REMARK 8.2 (The order of exponents). At first glance, the order of exponents in an expression like

$$e_{11} + e_{11}^3 + e_{11}^9 + e_{11}^5 + e_{11}^4$$

may look strange. In fact, it is natural: it is the geometric progression $1, 3, 9, 27, 81$ taken modulo 11. In the same way, the exponents in $e_7^3 + e_7^6 + e_7^5$ also constitute a progression $3, 6, 12$ modulo 7.

Whenever appropriate, we supply an additional information concerning the field in question, like the following one:

$$e_7^3 + e_7^6 + e_7^5 \in \mathbb{Q}(\sqrt{-7}).$$

An intimate relation of the irrationalities present in the character table with the fields of moduli literally jumps to the eyes, even if it is not every time direct and explicit. When there are several irrationalities in the character table, the one usually "chosen" for the field of moduli is the one which stays in the columns of the conjugacy classes corresponding to a partition present in the passport. (It is chosen not by us, of course, but by the very nature of things.) For example, for the Mathieu group M_{11}, among the two irrationalities $\sqrt{-2}$ and $\sqrt{-11}$ it is always the second one which is chosen "since" it stays in the columns of the classes of the cycle structure 11^1 for its action on 11 points, and of the cycle structure $11^1 1^1$ for its action on 12 points. (The word "since" in the previous sentence should not be understood as a theorem but as a mere observation.) For the Mathieu groups M_{23} and M_{24} what we see in the character tables is $\sqrt{-23}$ while the fields of moduli of the orbit 23.1 (with the monodromy group M_{23}) and of the orbit 24.4 (with the monodromy group M_{24}) are extensions of $\mathbb{Q}(\sqrt{-23})$.

The above observations should not be considered as an immutable law. Take, for example, the case 7.3: besides the orbit containing two special trees with the monodromy group $L_3(2)$ and defined over $\mathbb{Q}(\sqrt{-7})$, there are two generic trees with the monodromy group A_7 which are defined over $\mathbb{Q}(\sqrt{21})$. However, the character table of A_7 also contains $\sqrt{-7}$ but no $\sqrt{21}$. Should we be surprised? Probably, not. After all, the character tables of S_n contain only integers but the fields of moduli of maps with this monodromy group may contain arbitrary algebraic numbers.

Whenever the moduli field is known and its degree is greater than 2, we indicate also its Galois group.

We suggest to the reader to consult the proof of Proposition 10.5 (page 159). It follows that a tree with only one face of degree 1 is always irreducible (that is, its monodromy group is primitive), and a tree with two such faces may have, at most, a symmetry of order 2.

Tables and figures

AGL$_1$(5) of order 20 PrimitiveGroup(5,3)

Irrationalities in the character table: $e_4 = \sqrt{-1}$.

5.1. $(4^1 1^1, 2^2 1^1, 4^1 1^1)$. Number of trees: **2**.

$a = (2,3,4,5) \qquad b = (1,5)(2,3)$

Field of moduli: $\mathbb{Q}(\sqrt{-1})$.

There are no other trees with this passport.

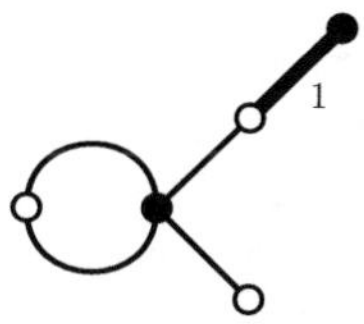

FIGURE 8.1. Group **AGL$_1$(5)**: orbit 5.1 of size 2.

L$_2$(5) of order 60 PrimitiveGroup(6,1)

Irrationalities in the character table: $e_5 + e_5^4 \in \mathbb{Q}(\sqrt{5})$.

6.1. $(3^2, 2^2 1^2, 5^1 1^1)$. Number of trees: **1**.

$a = (1,2,3)(4,5,6) \qquad b = (3,5)(4,6)$

Field of moduli: $\mathbb{Q}$. This is a unitree of the series D, see Chapter 5, page 31.

6.2. $(5^1 1^1, 2^2 1^2, 5^1 1^1)$. Number of trees: **1**.

$a = (1,5,3,2,4) \qquad b = (3,5)(4,6)$

Field of moduli: $\mathbb{Q}$.

There exist two more trees with this passport, defined over $\mathbb{Q}(\sqrt{-15})$.

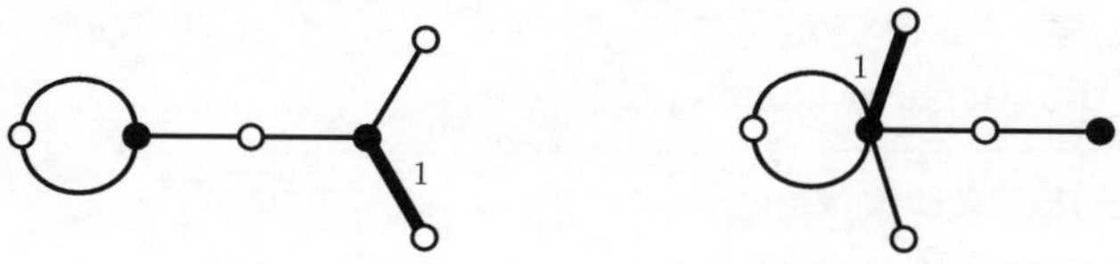

(A) Orbit 6.1. (B) Orbit 6.2.

FIGURE 8.2. Group **L$_2$(5)**: two orbits of size 1.

$PGL_2(5)$ of order 120 PrimitiveGroup(6,2)

No irrationalities in the character table.

6.3. $(4^1 1^2, 2^2 1^2, 6^1)$. Number of trees: **1**.

$a = (3,4,5,6) \qquad b = (1,5)(2,6)$

Field of moduli: $\mathbb{Q}$.

There exists one more tree with this passport, which is symmetric with the symmetry of order 2 and defined over $\mathbb{Q}$.

6.4. $(6^1, 2^2 1^2, 4^1 1^2)$. Number of trees: **1**.

$a = (1,2,3,5,4,6) \qquad b = (3,5)(4,6)$

Field of moduli: $\mathbb{Q}$. This tree is dual to the 6.3; therefore, the remarks to 6.3 remain valid also for 6.4.

6.5. $(4^1 1^2, 2^3, 5^1 1^1)$. Number of trees: **1**.

$a = (3,4,5,6) \qquad b = (1,3)(2,6)(4,5)$

Field of moduli: $\mathbb{Q}$. This is a unitree of the series J, see Chapter 5, page 33.

6.6. $(5^1 1^1, 2^3, 4^1 1^2)$. Number of trees: **1**.

$a = (2,3,4,6,5) \qquad b = (1,2)(3,4)(5,6)$

Field of moduli: $\mathbb{Q}$. This is a unitree of the series E_2, see Chapter 5, page 32.

6.7. $(4^1 1^2, 4^1 1^2, 5^1 1^1)$. Number of trees: **1**.

$a = (3,4,5,6) \qquad b = (1,3,6,2)$

Field of moduli: $\mathbb{Q}$. This is a unitree of the series C, see Chapter 5, page 31.

6.8. $(5^1 1^1, 4^1 1^2, 4^1 1^2)$. Number of trees: **1**.

$a = (1,2,4,3,6) \qquad b = (3,4,5,6)$

Field of moduli: $\mathbb{Q}$. This is a unitree of the series C, see Chapter 5, page 31.

6.9. $(4^1 1^2, 3^2, 4^1 1^2)$. Number of trees: **1**.

$a = (3,4,5,6) \qquad b = (1,3,2)(4,6,5)$

Field of moduli: $\mathbb{Q}$.

There exists one more tree with this passport, which is symmetric with the symmetry of order 2 and defined over $\mathbb{Q}$.

$AGL_1(7)$ of order 42 PrimitiveGroup(7,4)

Irrationalities in the character table: $e_3 \in \mathbb{Q}(\sqrt{-3})$.

7.1. $(3^2 1^1, 2^3 1^1, 6^1 1^1)$. Number of trees: **2**.

$a = (2,4,6)(3,5,7) \qquad b = (1,4)(3,5)(6,7)$

Field of moduli: $\mathbb{Q}(\sqrt{-3})$.

There are no other trees with this passport.

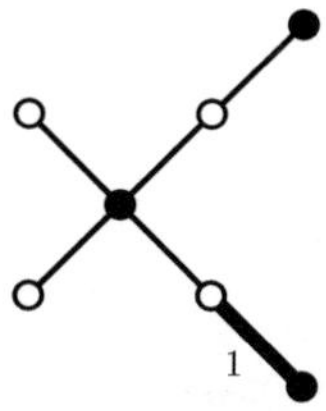

(A) Orbit 6.3.

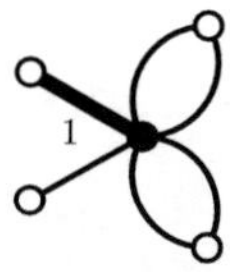

(B) Orbit 6.4.

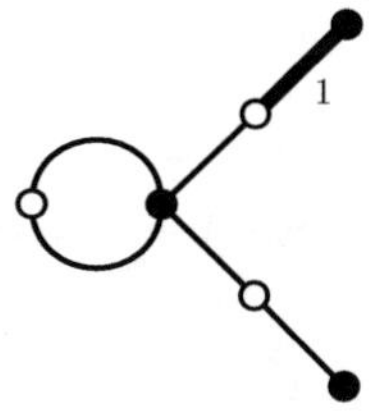

(C) Orbit 6.5.

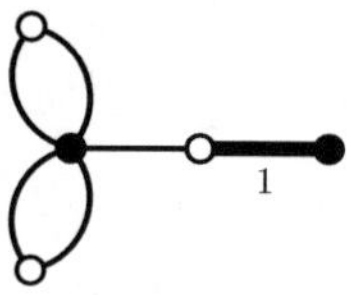

(D) Orbit 6.6.

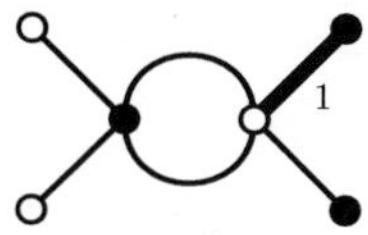

(E) Orbit 6.7.

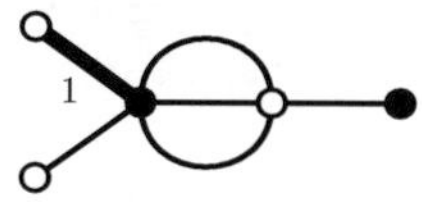

(F) Orbit 6.8.

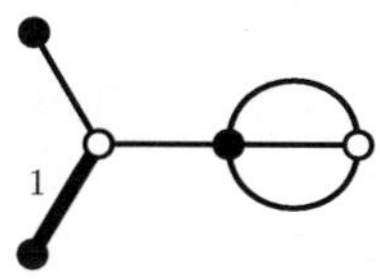

(G) Orbit 6.9.

FIGURE 8.3. Group $\mathbf{PGL}_2(5)$: seven orbits of size 1.

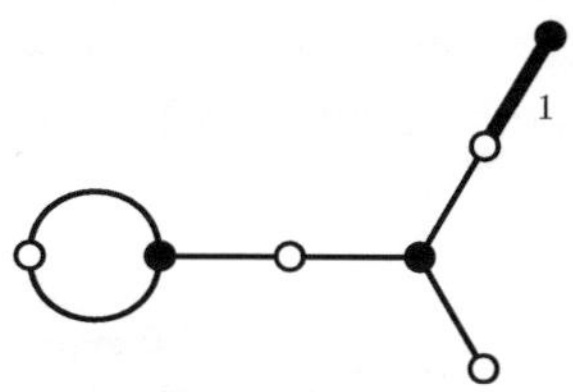

FIGURE 8.4. Group $\mathbf{AGL}_1(7)$: orbit 7.1 of size 2.

$L_3(2)$ of order 168 PrimitiveGroup(7,5)

Irrationalities in the character table: $e_7^3 + e_7^6 + e_7^5 \in \mathbb{Q}(\sqrt{-7})$.

7.2. $(3^2 1^1, 2^2 1^3, 7^1)$. Number of trees: **2**.

$$a = (2,3,5)(4,7,6) \qquad b = (1,3)(4,5)$$

Field of moduli: $\mathbb{Q}(\sqrt{-7})$.

There are no other trees with this passport.

7.3. $(4^1 2^1 1^1, 2^2 1^3, 7^1)$. Number of trees: **2**.

$$a = (2,3,4,7)(5,6) \qquad b = (1,4)(6,7)$$

Field of moduli: $\mathbb{Q}(\sqrt{-7})$.

There exist two more trees with this passport, defined over $\mathbb{Q}(\sqrt{21})$.

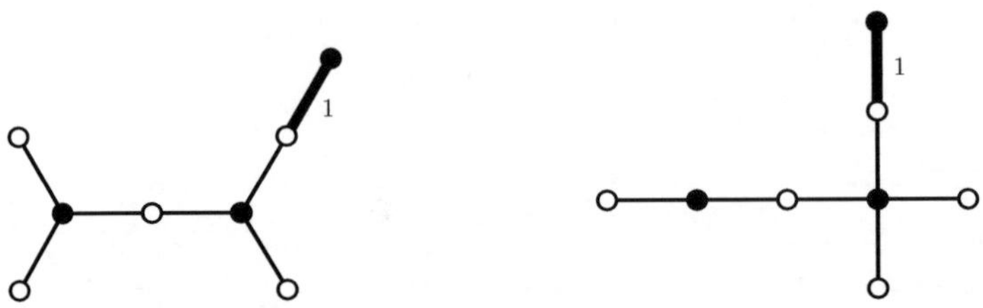

(A) Orbit 7.2. (B) Orbit 7.3.

FIGURE 8.5. Group $\mathbf{L}_3(2)$: two orbits of size 2.

$\mathbf{A\Gamma L}_1(8)$ of order 168 PrimitiveGroup(8,2)

Irrationalities in the character table: $e_3 \in \mathbb{Q}(\sqrt{-3})$, $e_7^3 + e_7^6 + e_7^5 \in \mathbb{Q}(\sqrt{-7})$.

8.1. $(3^2 1^2, 3^2 1^2, 7^1 1^1)$. Number of trees: **4**.

$$a_1 = (3,5,8)(4,6,7) \qquad b_1 = (1,4,8)(2,7,6)$$
$$a_2 = (1,4,8)(2,7,6) \qquad b_2 = (3,5,8)(4,6,7)$$

Field of moduli: $\mathbb{Q}(\sqrt{-3}, \sqrt{-7})$; its Galois group is Klein's group V_4 of order 4.

There exists one more tree with this passport, see orbit 8.9.

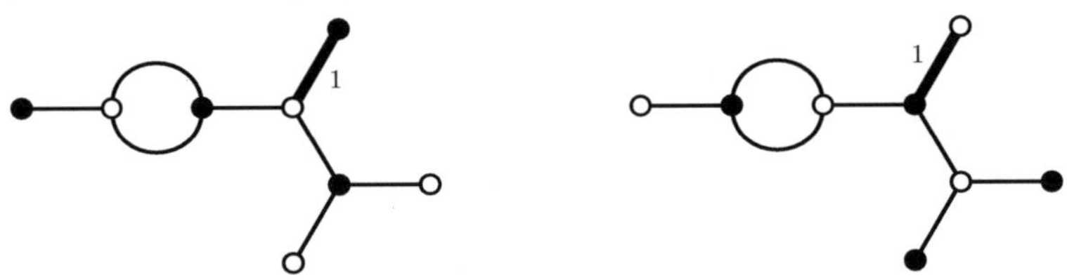

FIGURE 8.6. Group $\mathbf{A\Gamma L}_1(8)$: orbit 8.1 of size 4.

$\mathbf{ASL}_3(2)$ of order 1344 PrimitiveGroup(8,3)

Irrationalities in the character table: $e_7^3 + e_7^6 + e_7^5 \in \mathbb{Q}(\sqrt{-7})$.

8.2. $(4^1 2^1 1^2, 2^4, 7^1 1^1)$. Number of trees: **2**.

$$a = (3,4)(5,7,6,8) \qquad b = (1,3)(2,8)(4,6)(5,7)$$

Field of moduli: $\mathbb{Q}(\sqrt{-7})$.

There are no other trees with this passport.

8.3. $(4^2, 2^2 1^4, 7^1 1^1)$. Number of trees: **2**.

$$a = (1, 8, 4, 6)(2, 7, 3, 5) \qquad b = (2, 6)(4, 8)$$

Field of moduli: $\mathbb{Q}(\sqrt{-7})$.

There are no other trees with this passport.

8.4. $(6^1 2^1, 2^2 1^4, 7^1 1^1)$. Number of trees: **2**.

$$a = (1, 2)(3, 6, 7, 4, 5, 8) \qquad b = (2, 3)(6, 7)$$

Field of moduli: $\mathbb{Q}(\sqrt{-7})$.

There exist two more trees with this passport, defined over the field $\mathbb{Q}(\sqrt{-14})$.

8.5. $(7^1 1^1, 2^2 1^4, 7^1 1^1)$. Number of trees: **2**.

$$a = (2, 3, 5, 4, 7, 8, 6) \qquad b = (1, 3)(6, 8)$$

Field of moduli: $\mathbb{Q}(\sqrt{-7})$.

There exist three more trees with this passport, defined over the splitting field of the polynomial $a^3 - a^2 - 2a - 6$.

8.6. $(4^1 2^1 1^2, 3^2 1^2, 7^1 1^1)$. Number of trees: **4**.

$$a_1 = (1, 2, 7, 8)(4, 6) \qquad b_1 = (3, 5, 7)(4, 6, 8)$$
$$a_2 = (1, 8)(2, 4, 7, 5) \qquad b_2 = (3, 5, 7)(4, 6, 8)$$

Field of moduli: $\mathbb{Q}\left(\sqrt{-7}, \sqrt{2}\right)$; its Galois group is Klein's group V_4 of order 4.

There exist six more trees with this passport which form a single Galois orbit of degree 6.

8.7. $(4^1 2^1 1^2, 4^1 2^1 1^2, 7^1 1^1)$. Number of trees: **2**.

$$a = (3, 4)(5, 7, 6, 8) \qquad b = (1, 7)(2, 4, 8, 6)$$

Field of moduli: $\mathbb{Q}(\sqrt{-7})$.

There exist 18 more trees with this passport which form a single Galois orbit of degree 18.

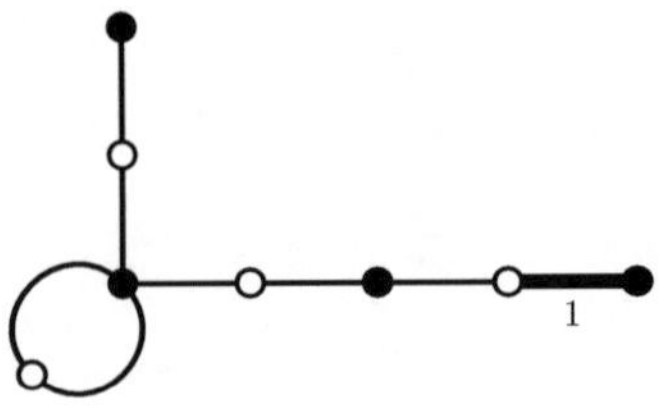

(A) Orbit 8.2.

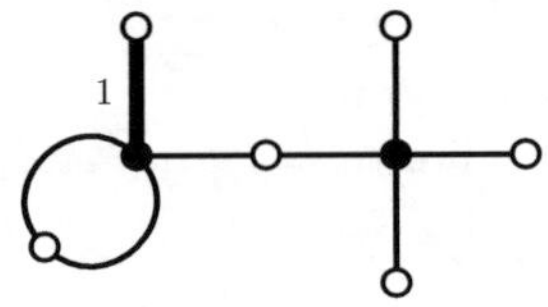

(B) Orbit 8.3.

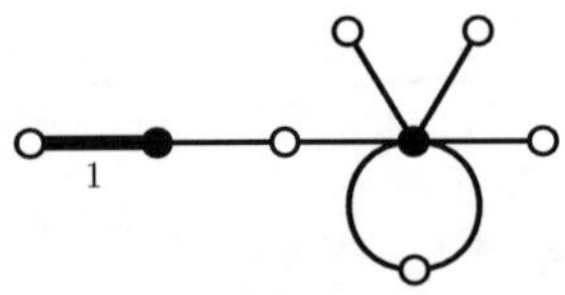

(C) Orbit 8.4.

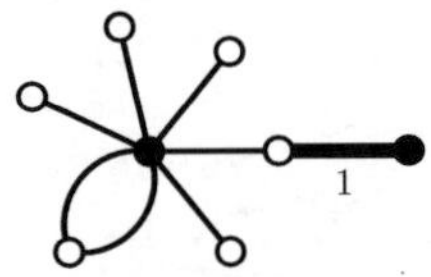

(D) Orbit 8.5.

(E) Orbit 8.6.

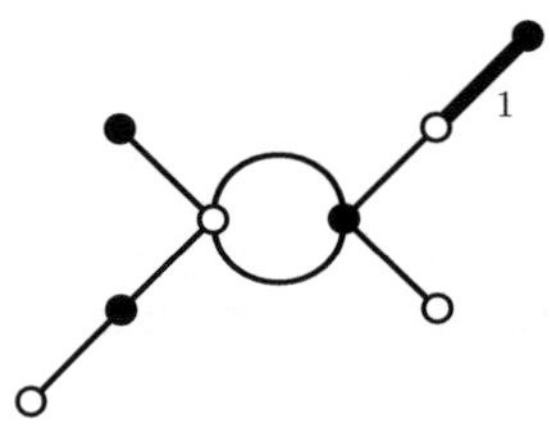

(F) Orbit 8.7.

FIGURE 8.7. Group $\mathbf{ASL}_3(2)$: five orbits of size 2 and one orbit of size 4.

$\mathbf{L}_2(7)$ of order 168 PrimitiveGroup(8,4)

Irrationalities in the character table: $e_7^3 + e_7^6 + e_7^5 \in \mathbb{Q}(\sqrt{-7})$.

8.8. $(3^21^2, 2^4, 7^11^1)$. Number of trees: **1**.

$a = (3,7,8)(4,6,5)$ $\quad$ $b = (1,3)(2,8)(4,5)(6,7)$

Field of moduli: $\mathbb{Q}$. This is the sporadic unitree K, see Chapter 5, page 33.

8.9. $(3^21^2, 3^21^2, 7^11^1)$. Number of trees: **1**.

$a = (3,7,8)(4,6,5)$ $\quad$ $b = (1,5,2)(4,8,6)$

Field of moduli: $\mathbb{Q}$.

There exist four more trees with this passport, see orbit 8.1.

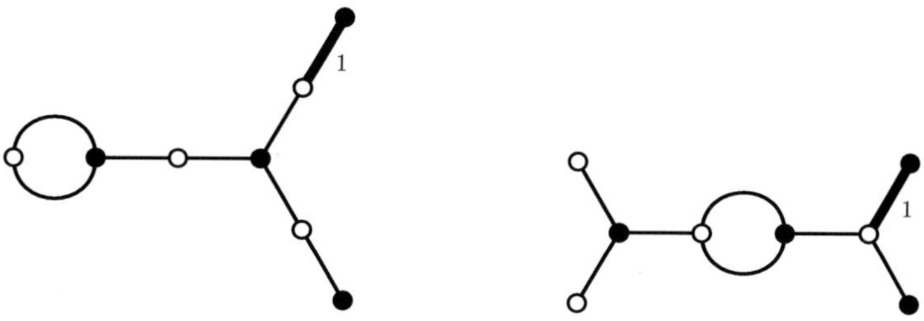

(A) Orbit 8.8. (B) Orbit 8.9.

FIGURE 8.8. Group $\mathbf{L}_2(7)$: two orbits of size 1.

$\mathbf{PGL}_2(7)$ of order 336 PrimitiveGroup(8,5)

Irrationalities in the character table: $-e_8 + e_8^3 \in \mathbb{Q}(\sqrt{2})$.

8.10. $(4^2, 2^3 1^2, 6^1 1^2)$. Number of trees: **1**.

$a = (1,2,3,8)(4,6,5,7) \qquad b = (2,6)(3,8)(4,7)$

Field of moduli: $\mathbb{Q}$.

There exist two more trees with this passport; they are both symmetric with the symmetry of order 2 and defined over $\mathbb{Q}(\sqrt{-2})$.

8.11. $(6^1 1^2, 2^4, 6^1 1^2)$. Number of trees: **1**.

$a = (3,5,7,4,8,6) \qquad b = (1,3)(2,6)(4,8)(5,7)$

Field of moduli: $\mathbb{Q}$.

There exists one more tree with this passport, which is symmetric with the symmetry of order 2 and defined over $\mathbb{Q}$.

8.12. $(6^1 1^2, 2^3 1^2, 7^1 1^1)$. Number of trees: **1**.

$a = (3,5,7,4,8,6) \qquad b = (1,6)(2,7)(4,8)$

Field of moduli: $\mathbb{Q}$.

There exist five more trees with this passport, defined over the splitting field of the polynomial $a^5 + 7a^3 - 14a^2 - 56$.

8.13. $(7^1 1^1, 2^3 1^2, 6^1 1^2)$. Number of trees: **1**.

$a = (2,3,8,6,7,5,4) \qquad b = (1,7)(4,5)(6,8)$

Field of moduli: $\mathbb{Q}$.

This tree is dual to 8.12; therefore, the remarks to 8.12 remain valid also for 8.13.

8.14. $(6^1 1^2, 3^2 1^2, 6^1 1^2)$. Number of trees: **1**.

$a = (3,5,7,4,8,6) \qquad b = (1,2,4)(3,6,5)$

Field of moduli: $\mathbb{Q}$.

There exist four more trees with this passport. One of them is symmetric, with the symmetry of order 2, and therefore it is defined over $\mathbb{Q}$. Three remaining trees are defined over the splitting field of the polynomial $a^3 - 6a + 16$.

8.15. $(3^2 1^2, 2^3 1^2, 8^1)$. Number of trees: **2**.

$a_1 = (3,7,8)(4,6,5) \qquad b_1 = (1,6)(2,7)(4,8)$

$a_2 = (3,7,8)(4,6,5) \qquad b_2 = (1,4)(2,6)(5,8)$

Field of moduli: $\mathbb{Q}(\sqrt{2})$.

There exist two more trees with this passport; they are both symmetric with the symmetry of order 2 and defined over $\mathbb{Q}(\sqrt{-2})$.

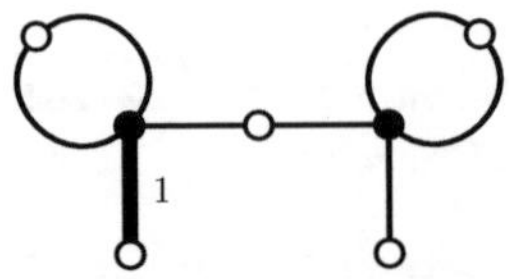

(A) Orbit 8.10.

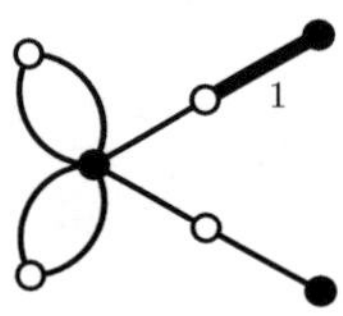

(B) Orbit 8.11.

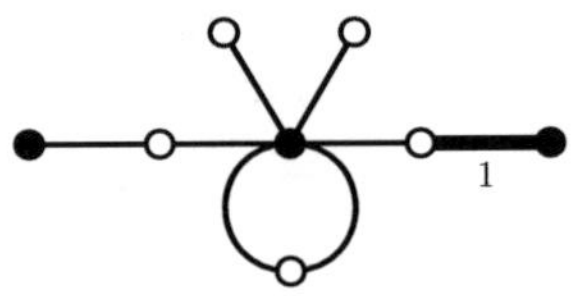

(C) Orbit 8.12.

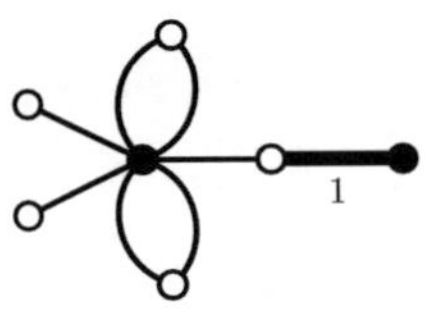

(D) Orbit 8.13.

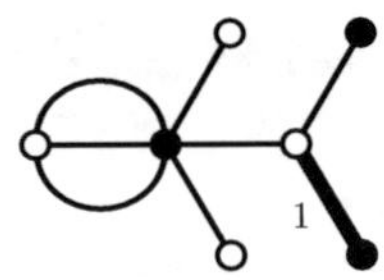

(E) Orbit 8.14.

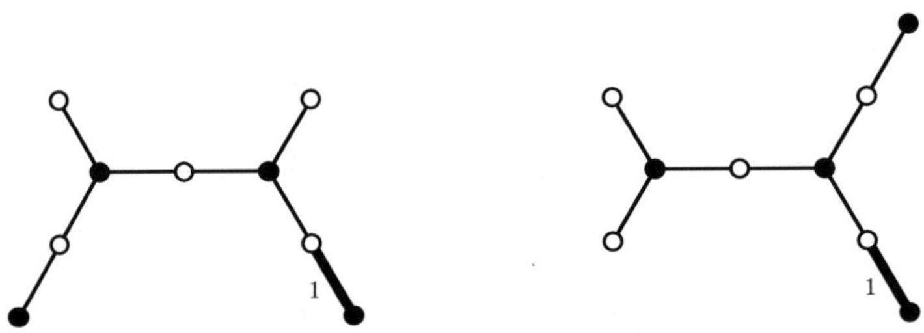

(F) Orbit 8.15.

FIGURE 8.9. Group $\mathbf{PGL}_2(7)$: five orbits of size 1 and one orbit of size 2.

AΓL$_1$(9) of order 144 PrimitiveGroup(9,5)

Irrationalities in the character table: $e_8 + e_8^3 = \sqrt{-2}$.

9.1. $(4^2 1^1, 2^3 1^3, 8^1 1^1)$. Number of trees: **2**.

$a = (2,4,3,7)(5,6,9,8) \qquad b = (1,3)(5,6)(7,8)$

Field of moduli: $\mathbb{Q}(\sqrt{-2})$.

There exist six more trees with this passport which form a single Galois orbit of degree 6.

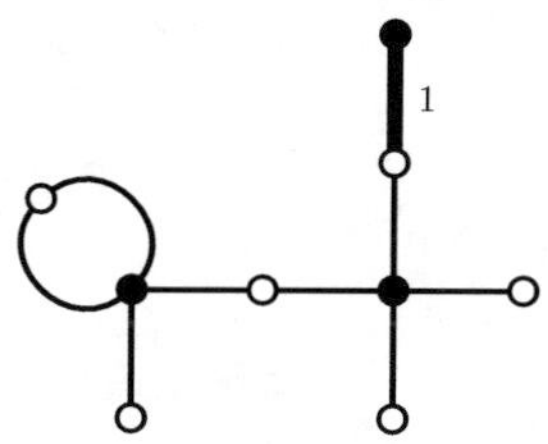

FIGURE 8.10. Group **AΓL$_1$(9)**: orbit 9.1 of size 2.

AGL$_2$(3) of order 432 PrimitiveGroup(9,7)

Irrationalities in the character table: $e_8 + e_8^3 = \sqrt{-2}$.

9.2. $(3^3, 2^3 1^3, 8^1 1^1)$. Number of trees: **2**.

$a = (1,2,4)(3,9,7)(5,6,8) \qquad b = (4,7)(5,8)(6,9)$

Field of moduli: $\mathbb{Q}(\sqrt{-2})$.

There are no other trees with this passport.

9.3. $(6^1 2^1 1^1, 2^3 1^3, 8^1 1^1)$. Number of trees: **2**.

$a = (2,3)(4,8,5,7,6,9) \qquad b = (1,7)(3,6)(5,8)$

Field of moduli: $\mathbb{Q}(\sqrt{-2})$.

There exist 14 more trees with this passport which form a single Galois orbit of degree 14.

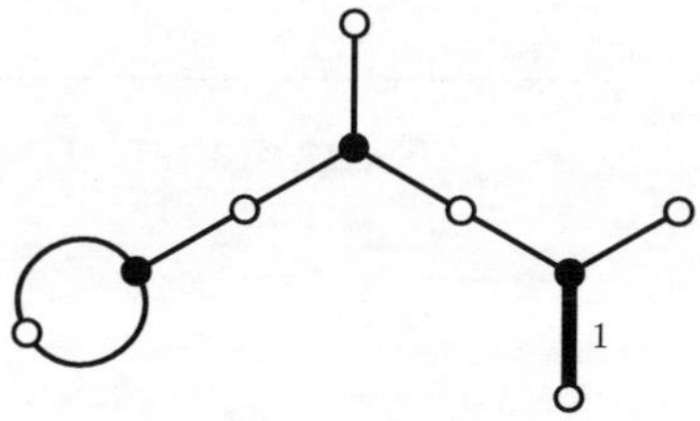

(A) Orbit 9.2.

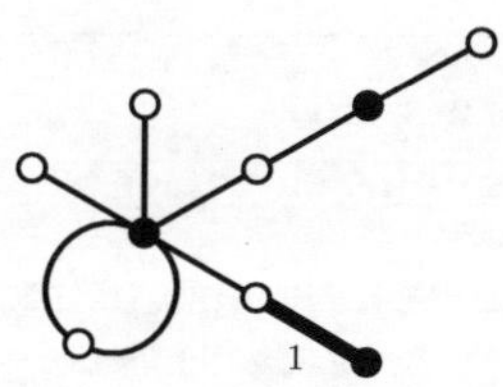

(B) Orbit 9.3.

FIGURE 8.11. Group **AGL$_2$(3)**: two orbits of size 2.

$\mathbf{L}_2(8)$ of order 504 PrimitiveGroup(9,8)

Irrationalities in the character table:
$e_7 + e_7^6, e_7^2 + e_7^5, e_7^3 + e_7^4$; these three numbers are the roots of $a^3 + a^2 - 2a - 1$;
$-e_9^2 - e_9^7, -e_9^4 - e_9^5, e_9^2 + e_9^4 + e_9^5 + e_9^7$; these three numbers are the roots of $a^3 - 3a - 1$.

9.4. $(3^3, 2^4 1^1, 7^1 1^2)$. Number of trees: **1**.

$$a = (1,2,5)(3,8,4)(6,9,7) \qquad b = (2,3)(4,8)(5,9)(6,7)$$

Field of moduli: $\mathbb{Q}$. This is the sporadic unitree L, see Chapter 5, page 33.

9.5. $(7^1 1^2, 2^4 1^1, 7^1 1^2)$. Number of trees: **2**.

$$a_1 = (3,4,9,8,6,7,5) \qquad b_1 = (1,6)(2,5)(3,4)(8,9)$$
$$a_2 = (3,4,9,8,6,7,5) \qquad b_2 = (1,9)(2,4)(3,5)(6,8)$$

Field of moduli: $\mathbb{Q}$ for both trees. This set consisting of two trees splits into two Galois orbits, both defined over $\mathbb{Q}$. We do not see a combinatorial (or any other) reason to explain this splitting.

There exist four more trees with this passport which form a single Galois orbit of degree 4.

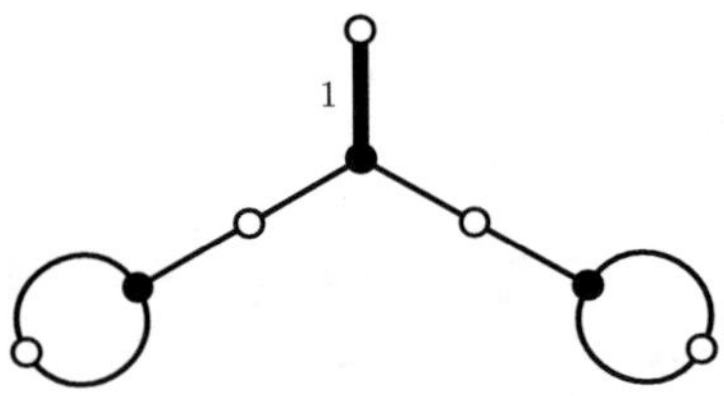

(A) Orbit 9.4.

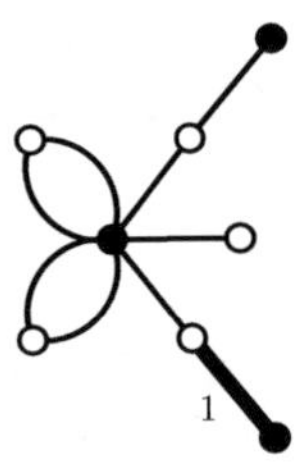

(B) Orbit 9.5(a).

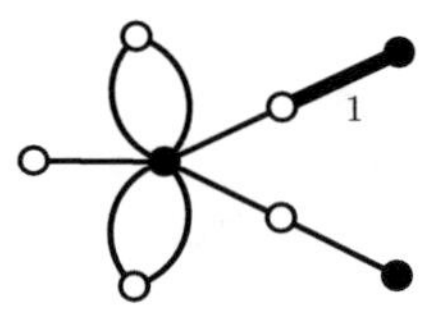

(C) Orbit 9.5(b).

FIGURE 8.12. Group $\mathbf{L}_2(8)$: three orbits of size 1.

$\mathbf{P\Gamma L}_2(8)$ of order 1512 PrimitiveGroup(9,9)

Irrationalities in the character table: $e_3 \in \mathbb{Q}(\sqrt{-3})$.

9.6. $(3^2 1^3, 2^4 1^1, 9^1)$. Number of trees: **2**.

$$a = (1,7,8)(3,4,9) \qquad b = (1,5)(2,8)(3,7)(6,9)$$

Field of moduli: $\mathbb{Q}(\sqrt{-3})$.

There are no other trees with this passport.

9.7. $(3^2 1^3, 3^2 1^3, 9^1)$. Number of trees: **4**.

$$a_1 = (1,7,8)(3,4,9) \qquad b_1 = (1,6,4)(2,8,5)$$
$$a_2 = (1,7,8)(3,4,9) \qquad b_2 = (1,4,6)(2,5,8)$$

Field of moduli: this combinatorial orbit consisting of four trees splits into two Galois orbits, both defined over $\mathbb{Q}(\sqrt{-3})$. There is a simple reason for this splitting: two trees are invariant under a color exchange, while the other two are not.

There are no other trees with this passport.

9.8. $(6^1 2^1 1^1, 3^2 1^3, 7^1 1^2)$. Number of trees: **4**.

$$a_1 = (1,3,7,4,8,9)(2,6) \qquad b_1 = (1,5,6)(4,9,8)$$
$$a_2 = (1,3,7,4,8,9)(2,6) \qquad b_2 = (1,9,5)(2,4,6)$$

Field of moduli: the splitting field of the polynomial $a^4 - 5a^2 + 43$, which is an extension of $\mathbb{Q}(\sqrt{-3})$. The Galois group of this field is the dihedral group D_4 of order 8.

There exist nine more trees with this passport which form a single Galois orbit of degree 9.

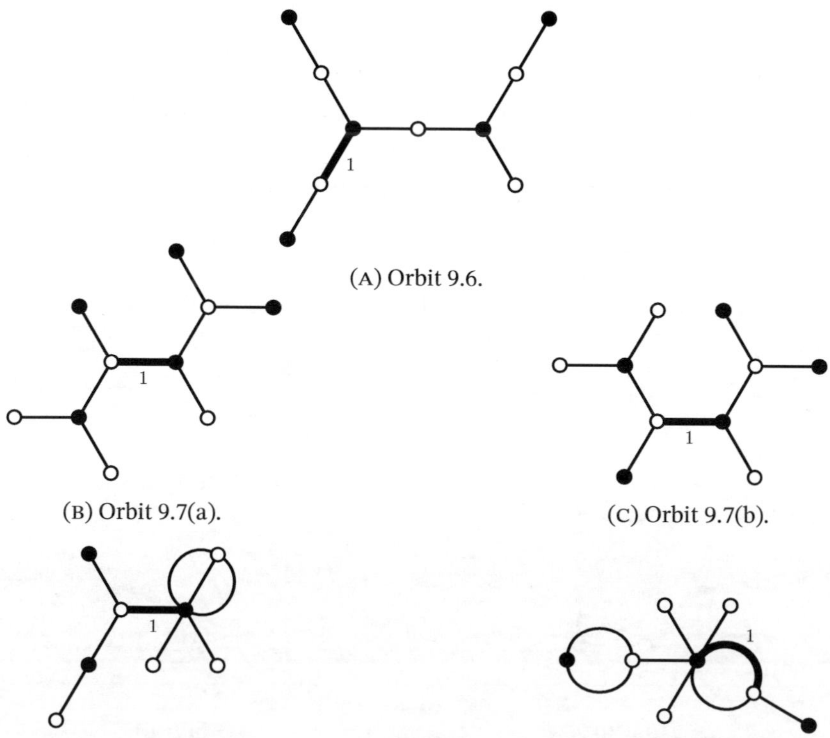

(A) Orbit 9.6.

(B) Orbit 9.7(a).

(C) Orbit 9.7(b).

(D) Orbit 9.8.

FIGURE 8.13. Group $\mathbf{P\Gamma L}_2(8)$: three orbits of size 2 and one orbit of size 4.

$\mathbf{PGL}_2(9)$ of order 720 PrimitiveGroup(10,4)

Irrationalities in the character table: $e_8 - e_8^3 = \sqrt{2}$; $e_5 + e_5^4 \in \mathbb{Q}(\sqrt{5})$

10.1. $(3^3 1^1, 2^5, 8^1 1^2)$. Number of trees: **1**.

$a = (2, 3, 10)(4, 9, 8)(5, 7, 6)$ $b = (1, 2)(3, 4)(5, 10)(6, 7)(8, 9)$

Field of moduli: $\mathbb{Q}$. This is the sporadic unitree M, see Chapter 5, page 33.

10.2. $(4^2 1^2, 2^5, 8^1 1^2)$. Number of trees: **1**.

$a = (3, 6, 10, 9)(4, 8, 5, 7)$ $b = (1, 3)(2, 7)(4, 8)(5, 6)(9, 10)$

Field of moduli: $\mathbb{Q}$.

There exist two more trees with this passport, which are symmetric with the symmetry of order 2 and defined over $\mathbb{Q}(\sqrt{-1})$.

10.3. $(8^1 1^2, 2^4 1^2, 8^1 1^2)$. Number of trees: **1**.

$a = (3, 4, 9, 7, 10, 5, 6, 8)$ $b = (1, 8)(2, 6)(4, 9)(7, 10)$

Field of moduli: $\mathbb{Q}$.

This example is rich with various combinatorial invariants. The *combinatorial* orbit corresponding to this passport consists of 16 trees, and it splits into *four* Galois orbits. One of them is our special tree; see Example 9.23.

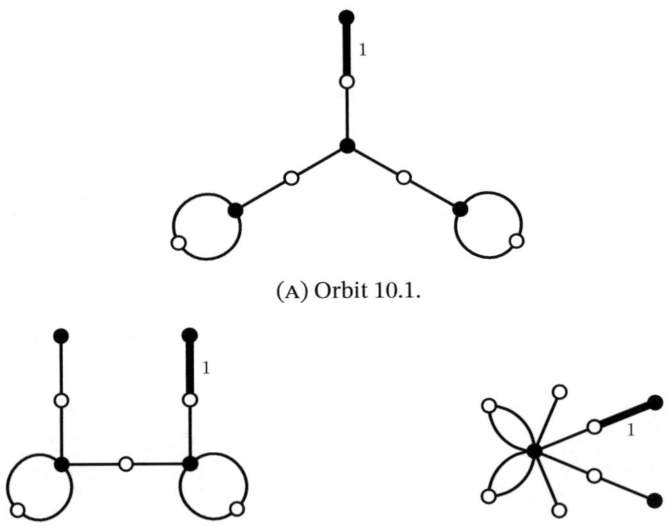

(A) Orbit 10.1.

(B) Orbit 10.2.

(C) Orbit 10.3.

FIGURE 8.14. Group $\mathbf{PGL}_2(9)$: three orbits of size 1.

PΓL$_2$(9) of order 1440 PrimitiveGroup(10,7)

No irrationalities in the character table.

10.4. $(8^1 2^1, 2^3 1^4, 8^1 1^2)$. Number of trees: **1**.

$a = (1, 2, 3, 6, 9, 8, 7, 4)(5, 10)$ $\qquad b = (1, 4)(2, 5)(3, 6)$

Field of moduli: $\mathbb{Q}$.

There exist nine more trees with this passport which form a single Galois orbit of degree 9.

10.5. $(4^2 1^2, 2^3 1^4, 10^1)$. Number of trees: **1**.

$a = (1, 5, 6, 9)(2, 10, 7, 8)$ $\qquad b = (1, 4)(2, 5)(3, 6)$

Field of moduli: $\mathbb{Q}$.

There exist eight more trees with this passport. Three of them are symmetric, with a symmetry of order 2; they are defined over the splitting field of the polynomial $a^3 - a^2 - 8a + 112$. The five remaining trees form a Galois orbit of degree 5.

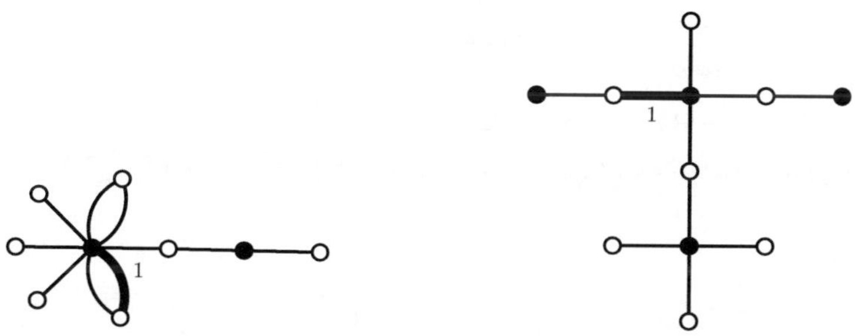

(A) Orbit 10.4. (B) Orbit 10.5.

FIGURE 8.15. Group **PΓL$_2$(9)**: two orbits of size 1.

L$_2$(11) of order 660 acting on 11 points PrimitiveGroup(11,5)

Irrationalities in the character table: $e_5^2 + e_5^3 \in \mathbb{Q}(\sqrt{5})$;
$e_{11} + e_{11}^3 + e_{11}^9 + e_{11}^5 + e_{11}^4 \in \mathbb{Q}(\sqrt{-11})$.

11.1. $(3^3 1^2, 2^4 1^3, 11^1)$. Number of trees: **2**.

$a = (3, 5, 11)(4, 9, 7)(6, 10, 8)$ $\qquad b = (1, 7)(2, 8)(5, 10)(6, 9)$

Field of moduli: $\mathbb{Q}(\sqrt{-11})$.

There exist eight more trees with this passport which form a single Galois orbit of degree 8.

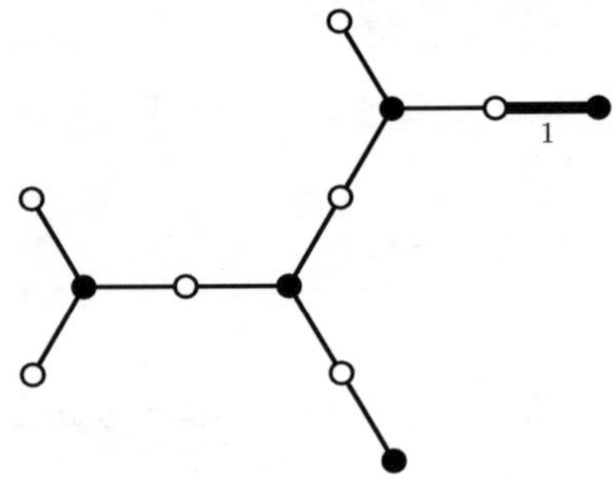

FIGURE 8.16. Group $\mathbf{L}_2(11)$ acting on 11 points: orbit 11.1 of size 2.

$\mathbf{M}_{11}$ of order 7920 acting on 11 points PrimitiveGroup(11,6)

Irrationalities in the character table: $e_8 + e_8^3 = \sqrt{-2}$;
$e_{11}^2 + e_{11}^6 + e_{11}^7 + e_{11}^{10} + e_{11}^8 \in \mathbb{Q}(\sqrt{-11})$.

11.2. $(4^21^3, 2^41^3, 11^1)$. Number of trees: **2**.

$a = (2,7,9,10)(3,4,8,5)$ $b = (1,7)(3,10)(5,11)(6,8)$

Field of moduli: $\mathbb{Q}(\sqrt{-11})$ (see [**Mat-96**]).

There exist eight more trees with this passport which form a single Galois orbit of degree 8.

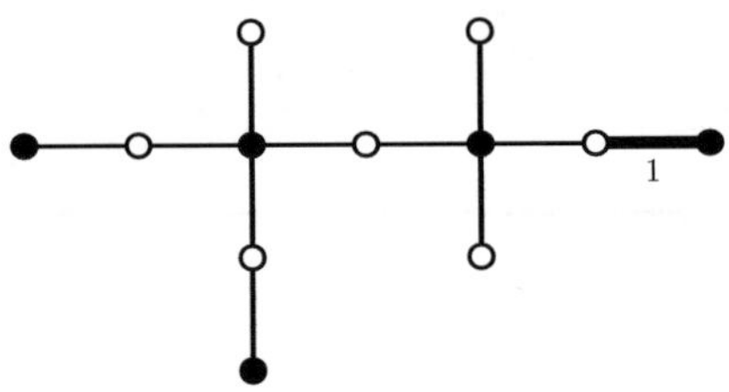

FIGURE 8.17. Group $\mathbf{M}_{11}$: orbit 11.2 of size 2.

$\mathbf{M}_{11}$ of order 7920 acting on 12 points PrimitiveGroup(12,1)

Irrationalities in the character table: $e_8 + e_8^3 = \sqrt{-2}$;
$e_{11}^2 + e_{11}^6 + e_{11}^7 + e_{11}^{10} + e_{11}^8 \in \mathbb{Q}(\sqrt{-11})$

12.1. $(4^22^2, 2^41^4, 11^11^1)$. Number of trees: **2**.

$a = (1,11,7,2)(3,5)(4,12,6,9)(8,10)$ $b = (1,4)(2,3)(5,10)(6,9)$

Field of moduli: $\mathbb{Q}(\sqrt{-11})$.

There exist 28 more trees with this passport which form a single Galois orbit of degree 28.

12.2. $(5^2 1^2, 2^4 1^4, 11^1 1^1)$. Number of trees: **2.**

$a = (1, 6, 11, 9, 8)(2, 12, 3, 5, 7)$ $\qquad b = (1, 10)(2, 12)(4, 7)(5, 11)$

Field of moduli: $\mathbb{Q}(\sqrt{-11})$.

There exist, in total, 45 trees with this passport. Two trees compose the orbit 12.2 with the monodromy group M_{11}; two more trees compose the orbit 12.7 with the monodromy group M_{12}; both orbits are defined over $\mathbb{Q}(\sqrt{-11})$. The 41 remaining trees form a single Galois orbit of degree 41.

12.3. $(6^1 3^1 2^1 1^1, 2^4 1^4, 11^1 1^1)$. Number of trees: **6.**

$a_1 = (1, 5, 7, 6, 8, 4)(2, 11, 10)(9, 12)$ $\qquad b_1 = (1, 10)(2, 11)(3, 6)(5, 12)$

$a_2 = (1, 5, 7, 6, 8, 4)(2, 11, 10)(9, 12)$ $\qquad b_2 = (1, 12)(2, 3)(6, 7)(8, 10)$

$a_3 = (1, 5, 7, 6, 8, 4)(2, 11, 10)(9, 12)$ $\qquad b_3 = (1, 3)(2, 12)(5, 9)(6, 7)$

Field of moduli: we believe that it is an extension of the field $\mathbb{Q}(\sqrt{-11})$.

There exist, in total, 150 trees with this passport. Six trees compose the orbit 12.3 with the monodromy group M_{11}; two more trees compose the orbit 12.8 with the monodromy group M_{12}. We can say nothing about the 142 remaining trees.

M_{12} of order 95 040 PrimitiveGroup(12,2)

Irrationalities in the character table: $e_{11}^2 + e_{11}^6 + e_{11}^7 + e_{11}^{10} + e_{11}^8 \in \mathbb{Q}(\sqrt{-11})$.

12.4. $(3^4, 2^4 1^4, 11^1 1^1)$. Number of trees: **2.**

$a = (1, 5, 6)(2, 8, 3)(4, 12, 7)(9, 11, 10)$ $\qquad b = (1, 8)(3, 9)(6, 12)(10, 11)$

Field of moduli: $\mathbb{Q}(\sqrt{-11})$.

There exist three more trees with this passport, defined over the splitting field of the polynomial $a^3 - a^2 + 4a + 2$.

12.5. $(3^3 1^3, 2^6, 11^1 1^1)$. Number of trees: **2.**

$a = (2, 7, 9)(3, 12, 10)(4, 11, 8)$ $\qquad b = (1, 11)(2, 7)(3, 6)(4, 5)(8, 10)(9, 12)$

Field of moduli: $\mathbb{Q}(\sqrt{-11})$.

There are no other trees with this passport.

12.6. $(4^2 1^4, 2^6, 11^1 1^1)$. Number of trees: **2.**

$a = (2, 10, 6, 4)(3, 9, 11, 7)$ $\qquad b = (1, 6)(2, 5)(3, 12)(4, 8)(7, 10)(9, 11)$

Field of moduli: $\mathbb{Q}(\sqrt{-11})$.

There are no other trees with this passport.

12.7. $(5^2 1^2, 2^4 1^4, 11^1 1^1)$. Number of trees: **2.**

$a = (1, 8, 9, 12, 2)(3, 5, 6, 4, 11)$ $\qquad b = (1, 10)(3, 5)(6, 12)(7, 9)$

Field of moduli: $\mathbb{Q}(\sqrt{-11})$.

See the comments to the orbit 12.2.

12.8. $(6^1 3^1 2^1 1^1, 2^4 1^4, 11^1 1^1)$. Number of trees: **2.**

$a = (2, 3, 9, 10, 7, 12)(4, 11, 8)(5, 6)$ $\qquad b = (1, 12)(3, 8)(6, 9)(7, 10)$

Field of moduli: $\mathbb{Q}(\sqrt{-11})$.

See the comments to the orbit 12.3.

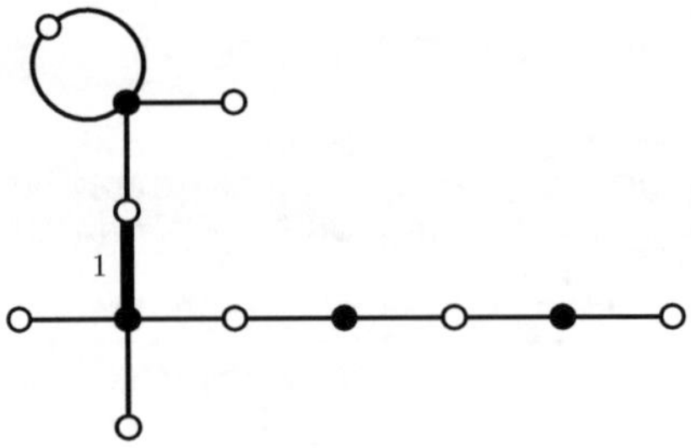

(A) Orbit 12.1.

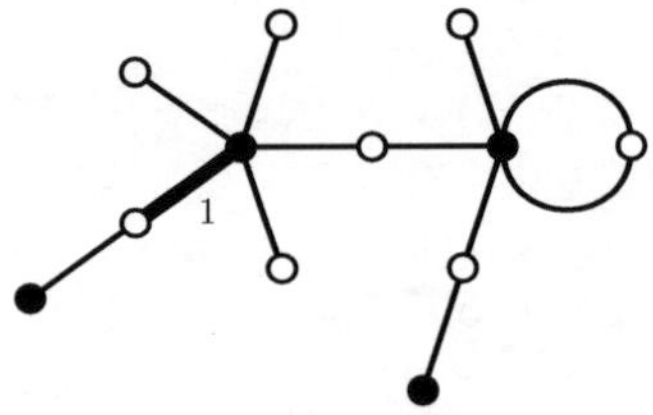

(B) Orbit 12.2.

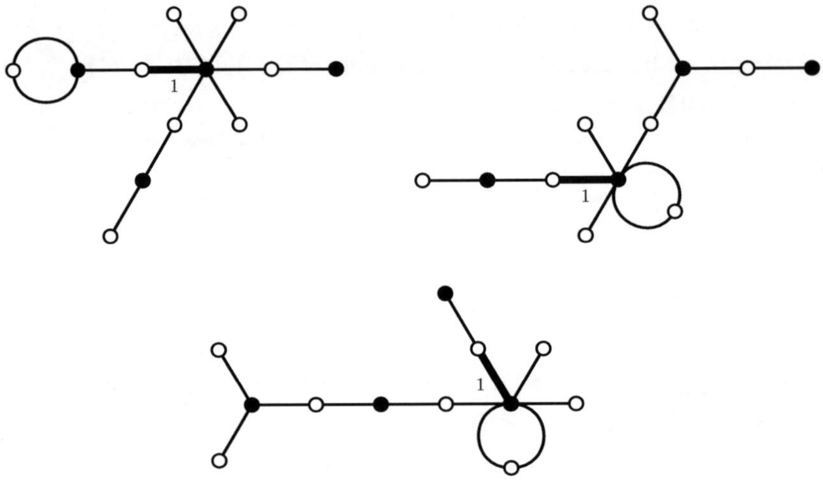

(C) Orbit 12.3.

FIGURE 8.18. Group $\mathbf{M}_{11}$ acting on 12 points: two orbits of size 2 and one orbit of size 6.

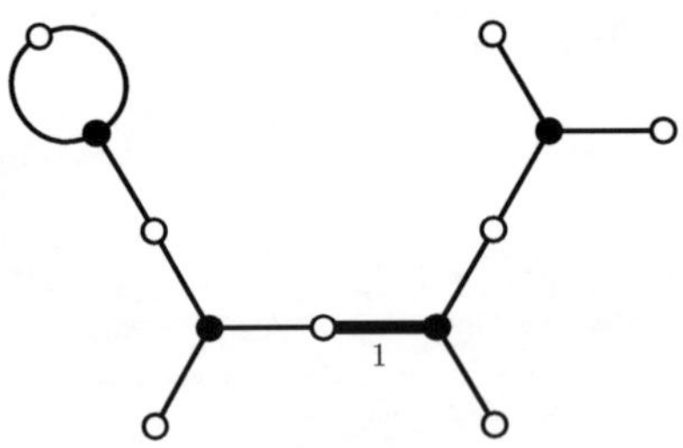

(A) Orbit 12.4.

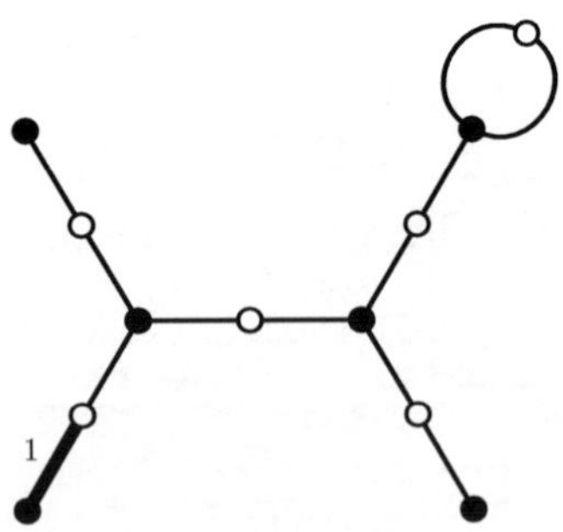

(B) Orbit 12.5.

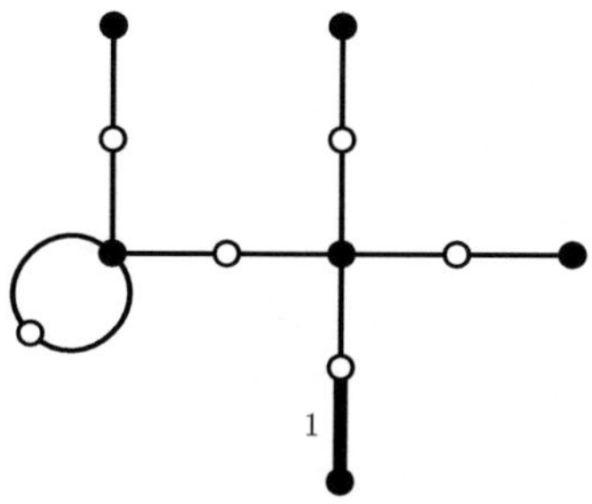

(C) Orbit 12.6.

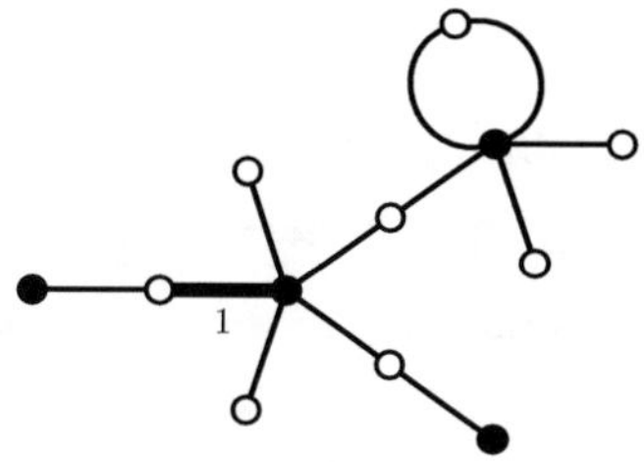

(D) Orbit 12.7.

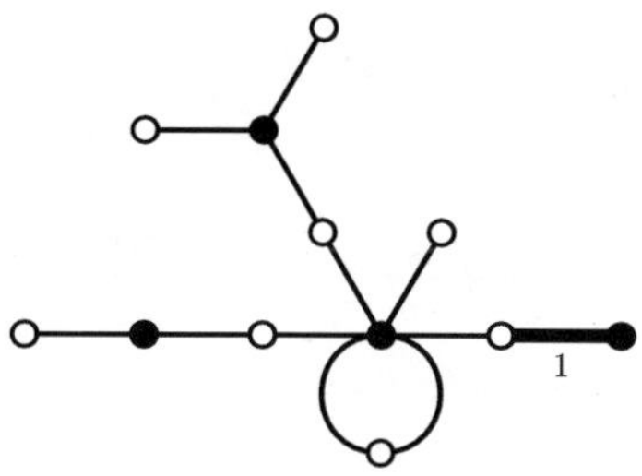

(E) Orbit 12.8.

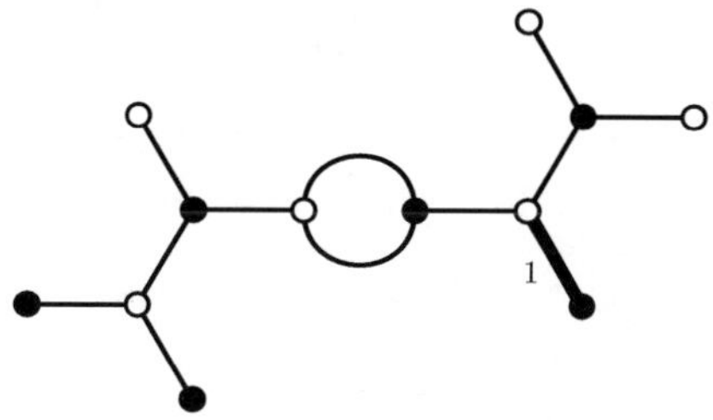

(G) Orbit 12.10.

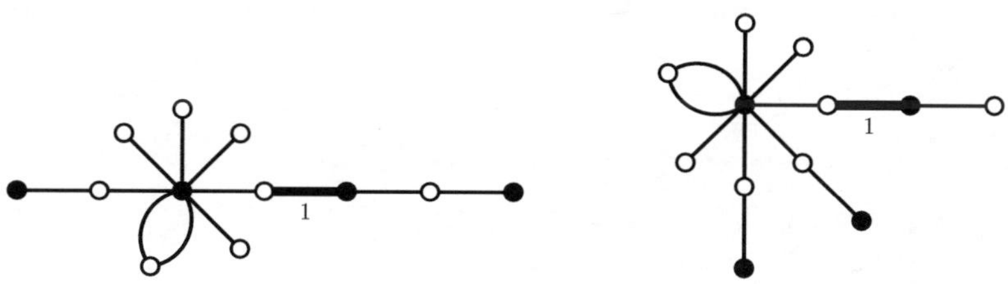

(F) Orbit 12.9.

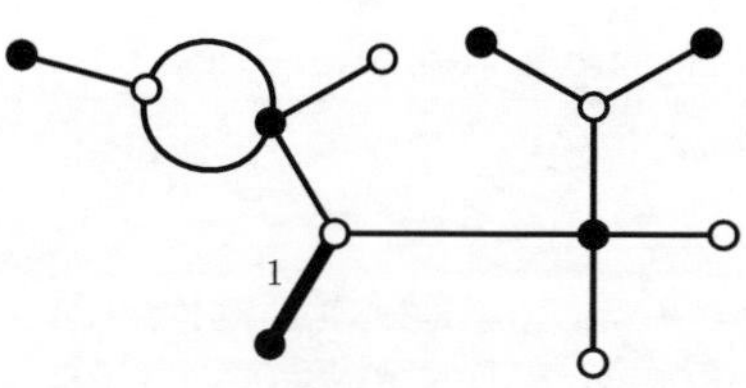

(H) Orbit 12.11.

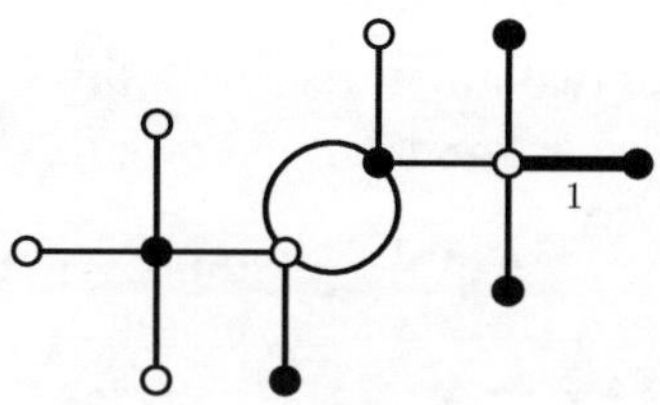

(I) Orbit 12.12.

FIGURE 8.20. Group $\mathbf{M}_{12}$: 8 orbits of size 2 and one orbit of size 4.

12.9. $(8^1 2^1 1^2, 2^4 1^4, 11^1 1^1)$. Number of trees: **4**.

$a_1 = (1,12)(2,3,10,9,6,11,4,7)$ $\quad$ $b_1 = (1,11)(3,8)(5,12)(9,10)$

$a_2 = (1,12)(2,3,10,9,6,11,4,7)$ $\quad$ $b_2 = (1,4)(3,10)(5,11)(6,8)$

Field of moduli: we believe that it is an extension of the field $\mathbb{Q}(\sqrt{-11})$.

There exist, in total, 90 trees with this passport.

12.10. $(3^3 1^3, 3^3 1^3, 11^1, 1^1)$. Number of trees: **2**.

$a = (2,7,9)(3,12,10)(4,11,8)$ $\quad$ $b = (1,9,10)(3,4,12)(5,8,6)$

Field of moduli: $\mathbb{Q}(\sqrt{-11})$.

There exist, in total, 36 trees with this passport.

12.11. $(4^2 1^4, 3^3 1^3, 11^1 1^1)$. Number of trees: **2**.

$a = (2,10,6,4)(3,9,11,7)$ $\quad$ $b = (1,6,11)(3,5,9)(8,10,12)$

Field of moduli: $\mathbb{Q}(\sqrt{-11})$.

There exist, in total, 26 trees with this passport.

12.12. $(4^2 1^4, 4^2 1^4, 11^1 1^1)$. Number of trees: **2**.

$a = (2,10,6,4)(3,9,11,7)$ $\quad$ $b = (1,12,11,5)(3,10,8,9)$

Field of moduli: $\mathbb{Q}(\sqrt{-11})$.

There exist, in total, 16 trees with this passport.

$PGL_2(11)$ of order 1320 PrimitiveGroup(12,4)

Irrationalities in the character table: $e_5^2 + e_5^3 \in \mathbb{Q}(\sqrt{5})$, $-e_{12}^7 + e_{12}^{11} = \sqrt{3}$

12.13. $(3^4, 2^5 1^2, 10^1 1^2)$. Number of trees: **2**.

$$a_1 = (1,7,8)(2,4,12)(3,6,11)(5,9,10)$$
$$b_1 = (2,9)(3,7)(5,10)(6,11)(8,12)$$
$$a_2 = (1,7,8)(2,4,12)(3,6,11)(5,9,10)$$
$$b_2 = (2,4)(3,11)(5,7)(6,9)(10,12)$$

Field of moduli: $\mathbb{Q}(\sqrt{5})$.

There exist two more trees with this passport; they are both symmetric with the symmetry of order 2 and defined over $\mathbb{Q}(\sqrt{-1})$.

12.14. $(5^2 1^2, 2^5 1^2, 10^1 1^2)$. Number of trees: **2**.

$a_1 = (2,12,9,7,4)(3,6,11,5,8)$ $\quad$ $b_1 = (1,7)(3,10)(4,8)(6,11)(9,12)$

$a_2 = (2,7,12,4,9)(3,5,6,8,11)$ $\quad$ $b_2 = (1,9)(3,4)(5,6)(8,11)(10,12)$

Field of moduli: $\mathbb{Q}(\sqrt{5})$.

There exist 34 more trees with this passport. Six of them are symmetric with the symmetry of order 2, but they split into two Galois orbits since two trees have the monodromy group of order 3840, while the other four have the group of order 23 040. The 28 remaining trees form a single Galois orbit of degree 28.

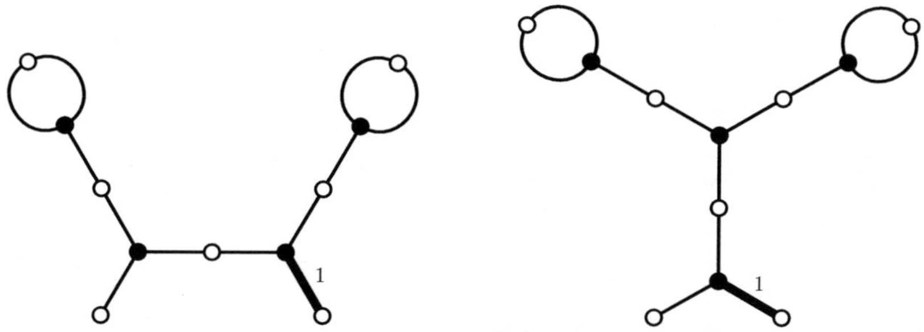

(A) Orbit 12.13.

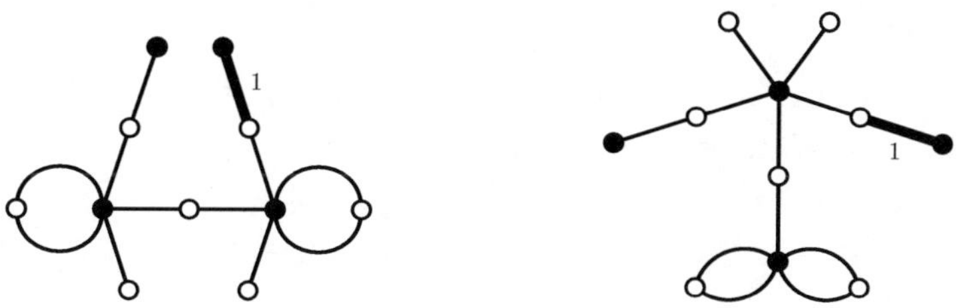

(B) Orbit 12.14.

FIGURE 8.21. Group $\mathbf{PGL}_2(11)$: two orbits of size 2.

$L_3(3)$ of order 5616 PrimitiveGroup(13,7)

Irrationalities in the character table: $e_8 + e_8^3 = \sqrt{-2}$;
$e_{13}^2 + e_{13}^6 + e_{13}^5$ and $e_{13}^4 + e_{13}^{12} + e_{13}^{10}$: these two numbers, together with their complex conjugates, are the roots of $P = a^4 + a^3 + 2a^2 - 4a + 3$. The Galois group of this polynomial is the cyclic group C_4 of order 4, and it can be factorized over the field $\mathbb{Q}(\sqrt{13})$:

$$P = \frac{1}{4} \cdot \left(2a^2 + (1+\sqrt{13})a + (5+\sqrt{13})\right) \cdot \left(2a^2 + (1-\sqrt{13})a + (5-\sqrt{13})\right).$$

13.1. $(3^4 1^1, 2^4 1^5, 13^1)$. Number of trees: **4**.

$$a_1 = (1,10,12)(2,9,7)(3,8,13)(4,5,11)$$
$$b_1 = (1,3)(2,12)(5,7)(6,10)$$
$$a_2 = (1,10,12)(2,9,7)(3,8,13)(4,5,11)$$
$$b_2 = (1,3)(2,8)(4,7)(6,11)$$

Field of moduli: the splitting field of $a^4 + a^3 + 2a^2 - 4a + 3$.

There exist, in total, 14 trees with this passport. The ten remaining trees form a single Galois orbit.

13.2. $(4^2 2^2 1^1, 2^4 1^5, 13^1)$. Number of trees: **4**.

$$a_1 = (1,11,3,8)(2,10,6,12)(4,13)(5,7)$$
$$b_1 = (2,4)(3,9)(5,8)(7,12)$$
$$a_2 = (1,11,3,8)(2,10,6,12)(4,13)(5,7)$$
$$b_2 = (2,4)(5,6)(9,13)(11,12)$$

Field of moduli: we believe that it is the splitting field of $a^4 + a^3 + 2a^2 - 4a + 3$.

There exist, in total, 84 trees with this passport.

13.3. $(6^1 3^1 2^1 1^2, 2^4 1^5, 13^1)$. Number of trees: **4**.

$$a_1 = (1,8,5)(2,7)(3,11,12,13,4,10)$$
$$b_1 = (1,9)(2,12)(4,5)(6,11)$$
$$a_2 = (1,8,5)(2,7)(3,11,12,13,4,10)$$
$$b_2 = (2,8)(3,9)(4,5)(6,10)$$

Field of moduli: we believe that it is the splitting field of $a^4 + a^3 + 2a^2 - 4a + 3$.

There exist, in total, 168 trees with this passport.

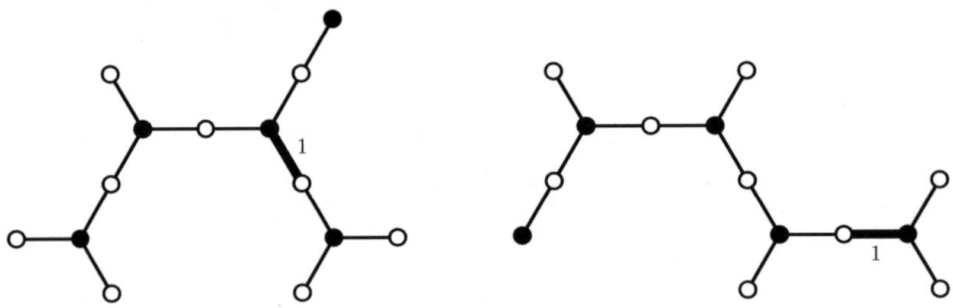

(A) Orbit 13.1.

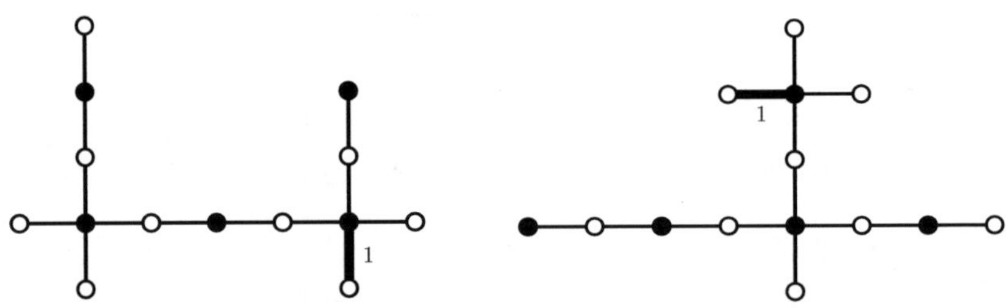

(B) Orbit 13.2.

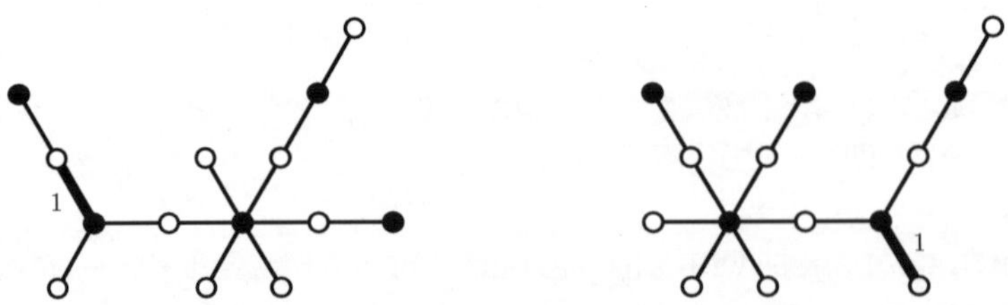

(C) Orbit 13.3.

FIGURE 8.22. Group $\mathbf{L}_3(3)$: three orbits of size 4.

$\mathbf{L}_2(13)$ of order 1092 PrimitiveGroup(14,1)

Irrationalities in the character table: $e_{13} + e_{13}^3 + e_{13}^9 + e_{13}^4 + e_{13}^{12} + e_{13}^{10} \in \mathbb{Q}(\sqrt{13})$;
$e_7 + e_7^6, e_7^2 + e_7^5, e_7^3 + e_7^4$: these three numbers are roots of $a^3 + a^2 - 2a - 1$.

14.1. $(3^4 1^2, 2^6 1^2, 13^1 1^1)$. Number of trees: **1**.

$$a = (2,14,12)(3,7,6)(4,9,10)(8,11,13)$$
$$b = (1,10)(2,12)(3,14)(4,6)(5,11)(7,8)$$

Field of moduli: $\mathbb{Q}$.

There exist, in total, 30 trees with this passport.

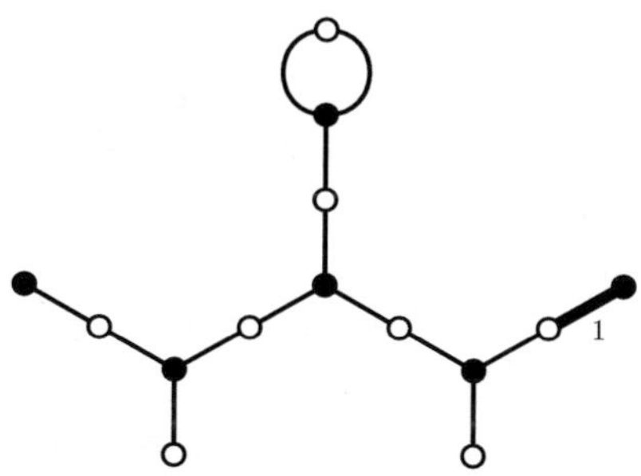

FIGURE 8.23. Group $\mathbf{L}_2(13)$: orbit 14.1 of size 1.

$\mathbf{PGL}_2(13)$ of order 2184 PrimitiveGroup(14,2)

Irrationalities in the character table: $-e_{12}^7 + e_{12}^{11} = \sqrt{3}$;
$e_7 + e_7^6, e_7^2 + e_7^5, e_7^3 + e_7^4$: these three numbers are roots of $a^3 + a^2 - 2a - 1$.

14.2. $(3^4 1^2, 2^7, 12^1 1^2)$. Number of trees: **2**.

$$a_1 = (1,7,14)(2,11,6)(3,13,4)(9,12,10)$$
$$b_1 = (1,7)(2,11)(3,9)(4,6)(5,10)(8,13)(12,14)$$
$$a_2 = (1,7,14)(2,11,6)(3,13,4)(9,12,10)$$
$$b_2 = (1,8)(2,12)(3,13)(4,10)(5,7)(6,11)(9,14)$$

Field of moduli: $\mathbb{Q}(\sqrt{3})$ (A. Vatuzov [**Vat-19**]).

There exist two more trees with this passport; they are both symmetric, with the symmetry of order 2.

14.3. $(4^3 1^2, 2^6 1^2, 12^1 1^2)$. Number of trees: **2**.

$$a_1 = (1,13,12,2)(3,9,6,14)(4,10,11,7)$$
$$b_1 = (1,13)(2,14)(3,10)(4,7)(5,12)(8,11)$$
$$a_2 = (1,13,12,2)(3,9,6,14)(4,10,11,7)$$
$$b_2 = (1,2)(3,9)(4,8)(5,7)(10,14)(11,13)$$

Field of moduli: we believe that it is $\mathbb{Q}(\sqrt{3})$.

There exist, in total, 62 trees with this passport. Two of them form the orbit 14.3; four trees are symmetric, with the symmetry of order 2; and there remain 56 generic trees.

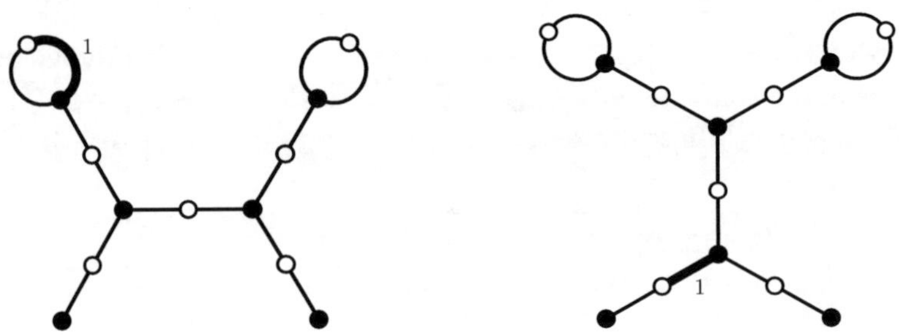

(A) Orbit 14.2.

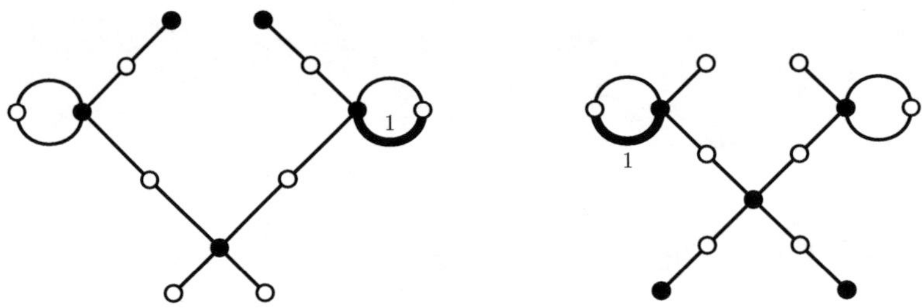

(B) Orbit 14.3.

FIGURE 8.24. Group $\mathbf{PGL}_2(13)$: two orbits of size 2.

$\mathbf{L}_4(2)$ of order 20 160 PrimitiveGroup(15,4)

Irrationalities in the character table: $e_7^3 + e_7^6 + e_7^5 \in \mathbb{Q}(\sqrt{-7})$;
$e_{15} + e_{15}^2 + e_{15}^4 + e_{15}^8 \in \mathbb{Q}(\sqrt{-15})$.

15.1. $(4^2 2^2 1^3, 2^6 1^3, 15^1)$. Number of trees: **2**.

$$a = (1,2,14,13)(3,12)(4,7,11,8)(6,9)$$
$$b = (1,7)(2,9)(3,14)(4,15)(5,8)(10,12)$$

Field of moduli: we believe that it is $\mathbb{Q}(\sqrt{-15})$.

There are, in total, 280 trees with this passport.

15.2. $(4^3 2^1 1^1, 2^4 1^7, 15^1)$. Number of trees: **2**.

$$a = (1,13,8,11)(2,7,4,14)(3,10,12,5)(6,9)$$
$$b = (2,3)(6,7)(10,11)(14,15)$$

Field of moduli: we believe that it is $\mathbb{Q}(\sqrt{-15})$.

There are, in total, 120 trees with this passport.

15.3. $(6^1 3^2 2^1 1^1, 2^4 1^7, 15^1)$. Number of trees: **2**.

$$a = (1,15,14)(2,5)(3,10,12,4,13,11)(6,8,9)$$
$$b = (1,2)(4,7)(9,10)(12,15)$$

Field of moduli: we believe that it is $\mathbb{Q}(\sqrt{-15})$.

There are, in total, 360 trees with this passport.

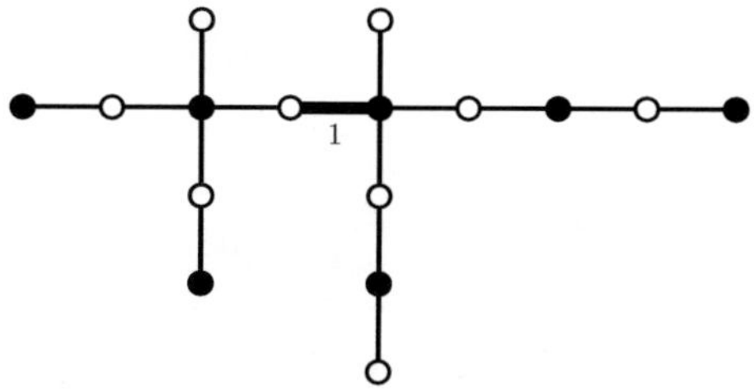

(A) Orbit 15.1.

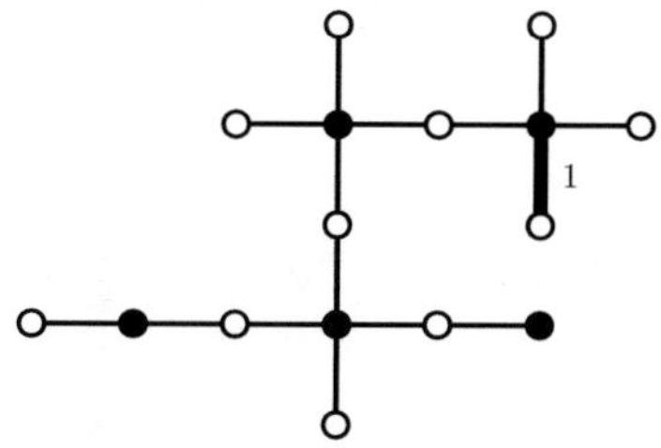

(B) Orbit 15.2.

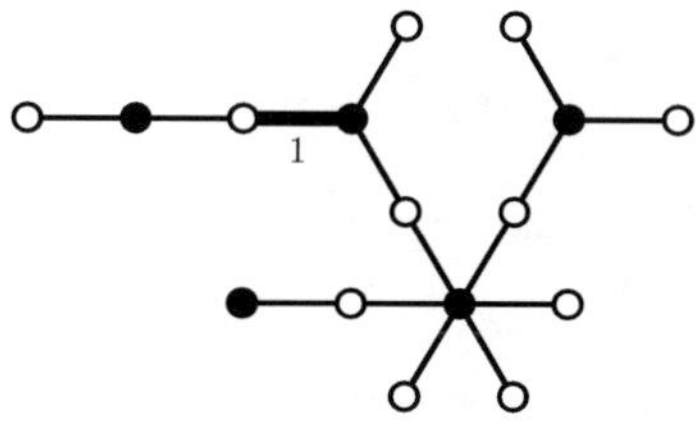

(C) Orbit 15.3.

FIGURE 8.25. Group $\mathbf{L}_4(2)$: three orbits of size 2.

$\mathbf{AGL}_4(2) = 2^4.\mathbf{L}_4(2)$ of order 322 560 PrimitiveGroup(16,11)

Irrationalities in the character table: $e_7^3 + e_7^6 + e_7^5 \in \mathbb{Q}(\sqrt{-7})$;
$e_{15} + e_{15}^2 + e_{15}^4 + e_{15}^8 \in \mathbb{Q}(\sqrt{-15})$.

16.1. $(4^2 2^4, 2^6 1^4, 15^1 1^1)$. Number of trees: **2**.

$$a = (1,7,16,10)(2,5)(3,9)(4,11,13,6)(8,14)(12,15)$$
$$b = (2,9)(4,11)(5,7)(6,15)(8,13)(14,16)$$

Field of moduli: we believe that it is $\mathbb{Q}(\sqrt{-15})$.

There are, in total, 140 trees with this passport.

16.2. $(5^3 1^1, 2^4 1^8, 15^1 1^1)$. Number of trees: **2**.

$$a = (2,11,4,16,8)(3,6,5,15,14)(7,12,10,13,9)$$
$$b = (1,15)(3,13)(5,11)(7,9)$$

Field of moduli: we believe that it is $\mathbb{Q}(\sqrt{-15})$.

There are, in total, 135 trees with this passport.

16.3. $(6^2 3^1 1^1, 2^4 1^8, 15^1 1^1)$. Number of trees: **2**.

$$a = (2,16,10,4,9,13)(3,8,6)(5,14,15,7,11,12)$$
$$b = (1,16)(3,14)(5,12)(7,10)$$

Field of moduli: we believe that it is $\mathbb{Q}(\sqrt{-15})$.

There are, in total, 405 trees with this passport.

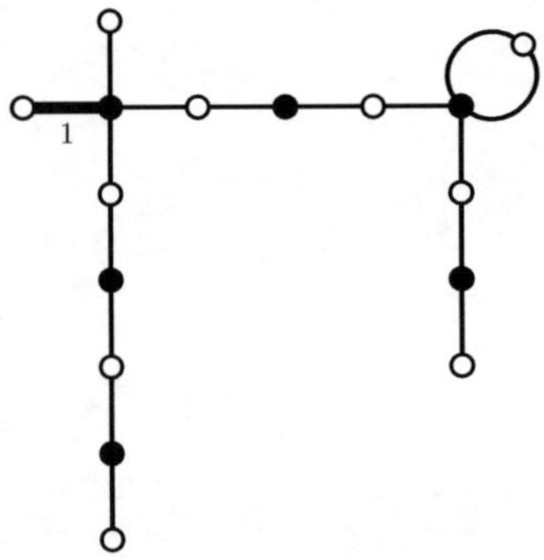

(A) Orbit 16.1.

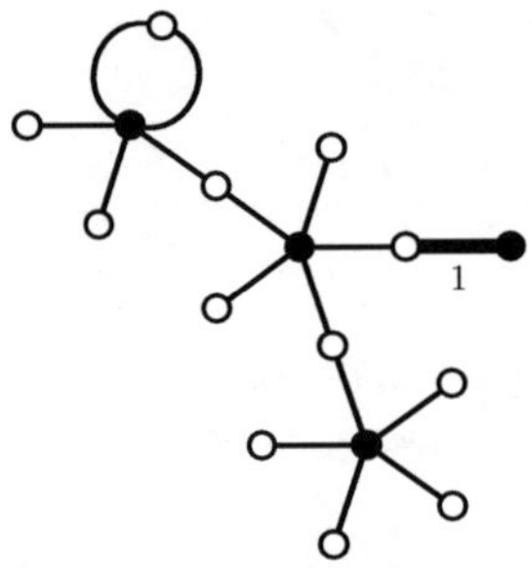

(B) Orbit 16.2.

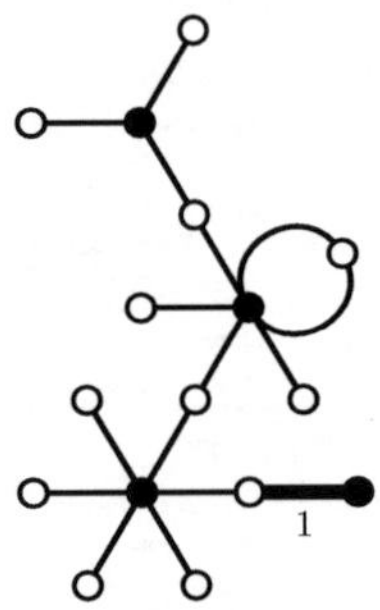

(C) Orbit 16.3.

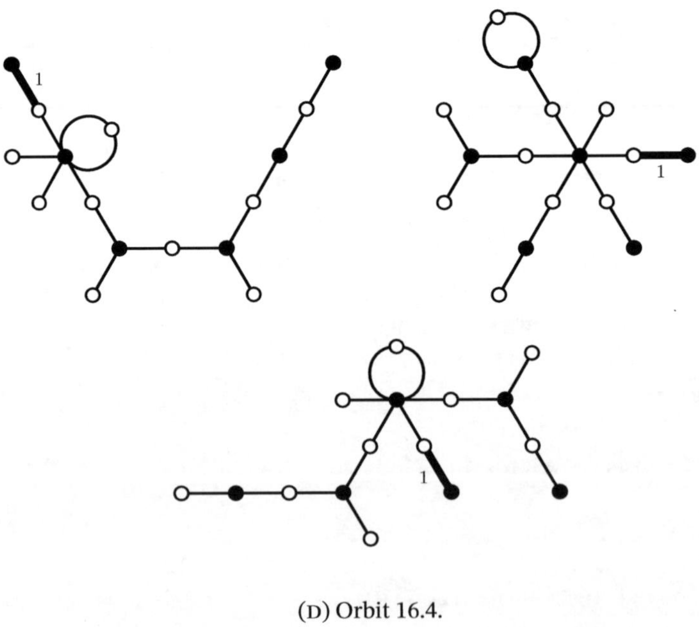

(D) Orbit 16.4.

FIGURE 8.26. Group $\mathbf{AGL}_4(2)$: three orbits of size 2 and one orbit of size 6.

16.4. $(6^1 3^2 2^1 1^2, 2^6 1^4, 15^1 1^1)$. Number of trees: **6.**

$$a_1 = (2,4,6,9,11,13)(3,7,14)(5,12,16)(8,15)$$
$$b_1 = (1,13)(3,15)(5,7)(6,12)(8,10)(9,11)$$
$$a_2 = (2,4,6,9,11,13)(3,7,14)(5,12,16)(8,15)$$
$$b_2 = (1,13)(4,16)(5,12)(6,7)(8,9)(10,11)$$
$$a_3 = (2,4,6,9,11,13)(3,7,14)(5,12,16)(8,15)$$
$$b_3 = (1,6)(4,7)(9,12)(10,16)(11,13)(14,15)$$

Field of moduli: we believe that it is an extension of the field $\mathbb{Q}(\sqrt{-15})$.

There are, in total, 2520 trees with this passport.

AΓL$_2$(4) of order 5760 PrimitiveGroup(16,12)

Irrationalities in the character table: $e_7^3 + e_7^6 + e_7^5 \in \mathbb{Q}(\sqrt{-7})$;
$e_{15} + e_{15}^2 + e_{15}^4 + e_{15}^8 \in \mathbb{Q}(\sqrt{-15})$.

16.5. $(4^3 2^1 1^2, 2^6 1^4, 15^1 1^1)$. Number of trees: **2.**

$$a = (3,16,6,10)(4,15,5,9)(7,8)(11,13,12,14)$$
$$b = (1,15)(2,12)(4,8)(5,11)(6,16)(10,14)$$

Field of moduli: we believe that it is $\mathbb{Q}(\sqrt{-15})$.

There are, in total, 840 trees with this passport.

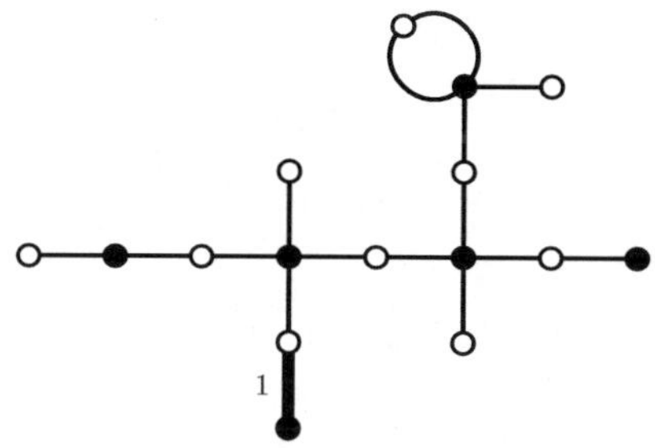

FIGURE 8.27. Group **AΓL$_2$(4)**: orbit 16.5 of size 2.

L$_2$(16) of order 4080 PrimitiveGroup(17,6)

Irrationalities in the character table: $e_5^2 + e_5^3 \in \mathbb{Q}(\sqrt{5})$;
$e_{15}+e_{15}^{14}, e_{15}^2+e_{15}^{13}, e_{15}^4+e_{15}^{11}, e_{15}^7+e_{15}^8$: these four numbers are the roots of the polynomial $P = a^4 - a^3 - 4a^2 + 4a + 1$ with the Galois group C_4 of order 4. The splitting field of this polynomial is a quadratic extension of the field $\mathbb{Q}(\sqrt{5})$. Indeed,

$$P = \frac{1}{4} \cdot (2a^2 - (1-\sqrt{5})a - (3+\sqrt{5}))(2a^2 - (1+\sqrt{5})a - (3-\sqrt{5})).$$

$e_{17} + e_{17}^{16}, e_{17}^2 + e_{17}^{15}, e_{17}^3 + e_{17}^{14}, e_{17}^4 + e_{17}^{13}, e_{17}^5 + e_{17}^{12}, e_{17}^6 + e_{17}^{11}, e_{17}^7 + e_{17}^{10}, e_{17}^8 + e_{17}^9$: these eight numbers are the roots of the polynomial

$$Q = a^8 + a^7 - 7a^6 - 6a^5 + 15a^4 + 10a^3 - 10a^2 - 4a + 1$$

with the Galois group C_8 of order 8. We also have

$$Q = \frac{1}{4} \cdot (2a^4 + (1+\sqrt{17})a^3 - (3-\sqrt{17})a^2 + (4-2\sqrt{17})a - 2)$$
$$\times(2a^4 + (1-\sqrt{17})a^3 - (3+\sqrt{17})a^2 + (4+2\sqrt{17})a - 2).$$

17.1. $(3^5 1^2, 2^8 1^1, 15^1 1^2)$. Number of trees: **1**.

$a = (1,14,11)(2,7,15)(3,8,12)(4,6,5)(9,16,13)$
$b = (1,11)(2,10)(3,9)(4,8)(5,7)(12,17)(13,16)(14,15)$

Field of moduli: $\mathbb{Q}$.

There are, in total, 30 trees with this passport.

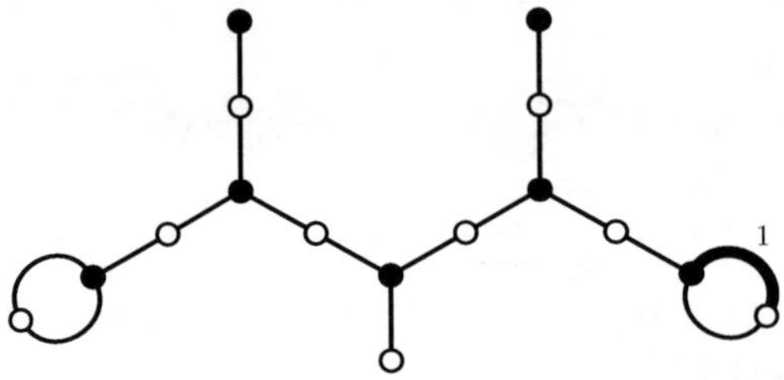

FIGURE 8.28. Group $\mathbf{L}_2(16)$: orbit 17.1 of size 1.

$\mathbf{L}_2(16):2$ of order 8160 PrimitiveGroup(17,7)

Irrationalities in the character table: $e_5^2 + e_5^3 \in \mathbb{Q}(\sqrt{5})$;
$e_{17} + e_{17}^4 + e_{17}^{16} + e_{17}^{13}$, $e_{17}^2 + e_{17}^8 + e_{17}^{15} + e_{17}^9$, $e_{17}^3 + e_{17}^{12} + e_{17}^{14} + e_{17}^5$, $e_{17}^6 + e_{17}^7 + e_{17}^{11} + e_{17}^{10}$: these four numbers are the roots of the polynomial $a^4 + a^3 - 6a^2 - a + 1$ with the Galois group C_4 of order 4. We also have

$$a^4 + a^3 - 6a^2 - a + 1 = \frac{1}{4} \cdot (2a^2 + (1+\sqrt{17})a - 2)(2a^2 + (1-\sqrt{17})a - 2).$$

17.2. $(4^4 1^1, 2^6 1^5, 15^1 1^2)$. Number of trees: **1**.

$a = (2,7,10,6)(3,12,15,13)(4,11,16,5)(8,17,14,9)$
$b = (1,13)(2,6)(3,10)(4,11)(8,12)(15,16)$

Field of moduli: $\mathbb{Q}$.

There are, in total, 336 trees with this passport.

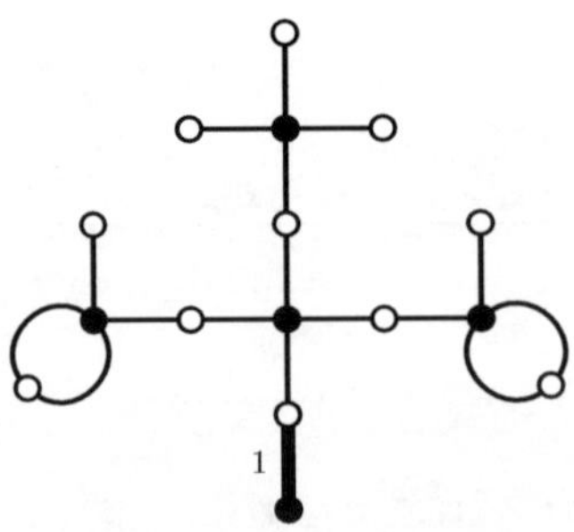

FIGURE 8.29. Group $\mathbf{L}_2(16):2$: orbit 17.2 of size 1.

$PGL_2(19)$ of order 6840 PrimitiveGroup(20,2)

Irrationalities in the character table: $e_5 + e_5^4 \in \mathbb{Q}(\sqrt{5})$;
$e_9^2 + e_9^7, e_9^4 + e_9^5, -e_9^2 - e_9^4 - e_9^5 - e_9^7$: these three numbers are roots of $a^3 - 3a + 1$ (Galois group C_3); $b = -e_{20} + e_{20}^9, c = -e_{20}^{13} + e_{20}^{17}$: these two numbers, together with $-b$ and $-c$, are the roots of the biquadratic polynomial $a^4 - 5a^2 + 5$ (Galois group C_4); its splitting field is a quadratic extension of $\mathbb{Q}(\sqrt{5})$.

20.1. $(3^6 1^2, 2^9 1^2, 18^1 1^2)$. Number of trees: **3**.

$a_1 = (1,8,5)(3,4,17)(6,9,19)(7,15,14)(10,16,12)(11,18,20)$
$b_1 = (1,8)(2,12)(4,18)(5,7)(6,9)(10,19)(11,14)(13,15)(16,20)$

$a_2 = (1,8,5)(3,4,17)(6,9,19)(7,15,14)(10,16,12)(11,18,20)$
$b_2 = (1,9)(2,20)(4,14)(5,10)(6,19)(7,15)(11,17)(12,18)(13,16)$

$a_3 = (1,8,5)(3,4,17)(6,9,19)(7,15,14)(10,16,12)(11,18,20)$
$b_3 = (1,16)(2,6)(4,12)(5,9)(7,14)(8,15)(10,18)(11,20)(13,17)$

Field of moduli: we believe that it is the splitting field of $a^3 - 3a + 1$.

There exist, in total, 216 trees with this passport; three of them form the orbit 20.1, and 12 trees are symmetric, with the symmetry of order 2.

$P\Gamma L_3(4)$ of order 120 960 PrimitiveGroup(21,7)

Irrationalities in the character table: $e_7^3 + e_7^6 + e_7^5 \in \mathbb{Q}(\sqrt{-7})$,
$e_{15} + e_{15}^2 + e_{15}^4 + e_{15}^8 \in \mathbb{Q}(\sqrt{-15})$.

21.1. $(4^4 2^1 1^3, 2^7 1^7, 21^1)$. Number of trees: **2**.

$a = (2,13,16,11)(3,10,9,7)(4,15,20,12)(5,19)(6,14,8,17)$
$b = (1,16)(4,17)(5,20)(6,18)(10,11)(12,13)(14,21)$

Field of moduli: $\mathbb{Q}(\sqrt{-7})$ (see [**CaCo-99**]).

There are, in total, 8580 trees with this passport.

M_{23} of order 10 200 960 PrimitiveGroup(23,5)

Irrationalities in the character table: $e_7^3 + e_7^6 + e_7^5 \in \mathbb{Q}(\sqrt{-7})$;
$e_{11}^2 + e_{11}^6 + e_{11}^7 + e_{11}^{10} + e_{11}^8 \in \mathbb{Q}(\sqrt{-11})$; $e_{15} + e_{15}^2 + e_{15}^4 + e_{15}^8 \in \mathbb{Q}(\sqrt{-15})$;
$e_{23}^5 + e_{23}^{10} + e_{23}^{20} + e_{23}^{17} + e_{23}^{11} + e_{23}^{22} + e_{23}^{21} + e_{23}^{19} + e_{23}^{15} + e_{23}^7 + e_{23}^{14} \in \mathbb{Q}(\sqrt{-23})$.

23.1. $(4^4 2^2 1^3, 2^8 1^7, 23^1)$. Number of trees: **4**.

$a_1 = (1,14,17,12)(3,20,7,6)(5,9,16,22)(8,15,10,23)(11,21)(13,18)$
$b_1 = (2,5)(3,21)(4,17)(8,16)(10,18)(11,19)(12,23)(14,20)$

$a_2 = (1,14,17,12)(3,20,7,6)(5,9,16,22)(8,15,10,23)(11,21)(13,18)$
$b_2 = (2,6)(3,13)(4,17)(8,9)(10,12)(11,18)(15,20)(19,23)$

Field of moduli: splitting field of $a^4 + 23a^2 + 276$ (Matiyasevich [**Mat-98**]), or, alternatively, of $a^4 - 46a^2 + 621$ (Elkies [**Elk-13**], Müller [**Mul-15**]). Both presentations give the same field, which is an extension of $\mathbb{Q}(\sqrt{-23})$. The Galois group of this field is the dihedral group D_4 of order 8.

There are, in total, 60 060 trees with this passport.

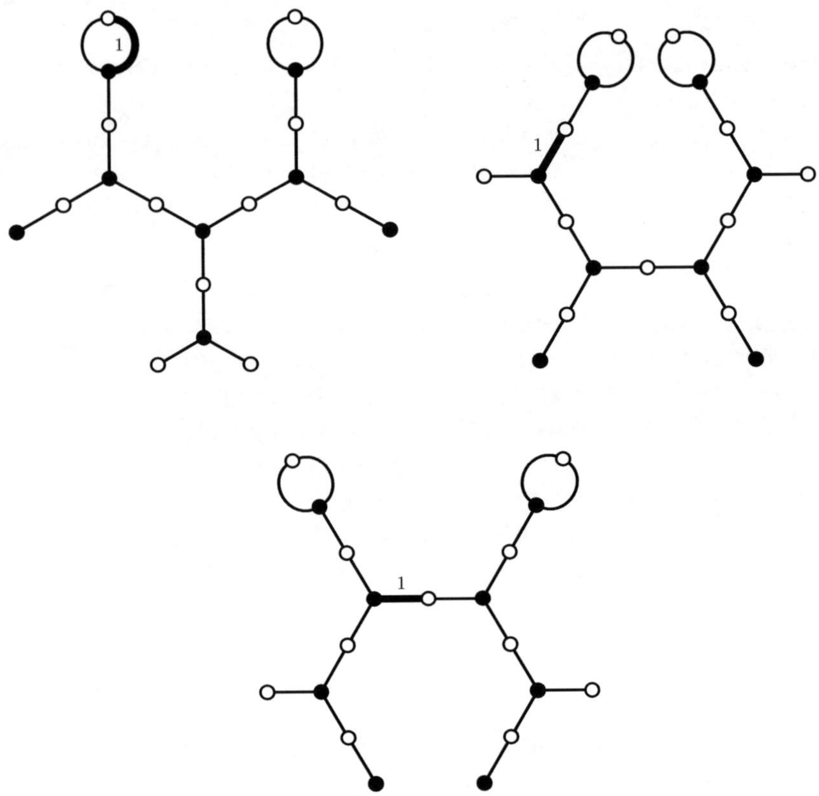

FIGURE 8.30. Group $\mathbf{PGL}_2(19)$: orbit 20.1 of size 3.

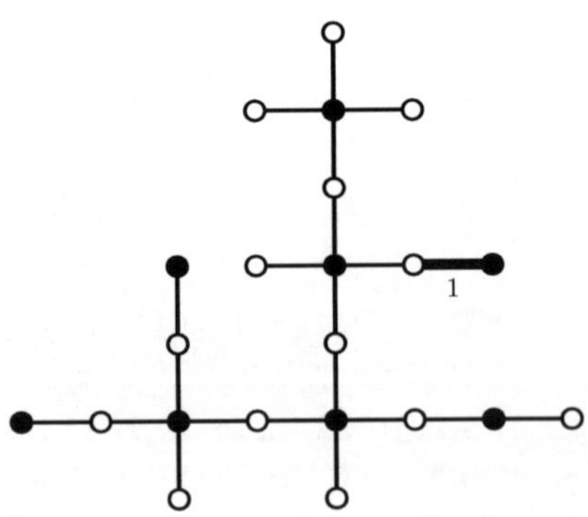

FIGURE 8.31. Group $\mathbf{P\Gamma L}_3(4)$: orbit 21.1 of size 2.

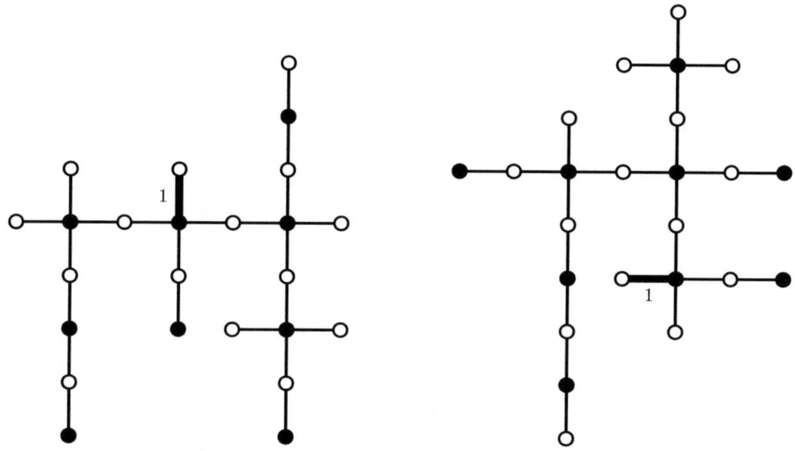

FIGURE 8.32. Group $\mathbf{M}_{23}$: orbit 23.1 of size 4.

$\mathbf{M}_{24}$ of order 244 823 040 PrimitiveGroup(24,1)

Irrationalities in the character table: $e_7^3 + e_7^6 + e_7^5 \in \mathbb{Q}(\sqrt{-7})$;
$e_{15} + e_{15}^2 + e_{15}^4 + e_{15}^8 \in \mathbb{Q}(\sqrt{-15})$;
$e_{23}^5 + e_{23}^{10} + e_{23}^{20} + e_{23}^{17} + e_{23}^{11} + e_{23}^{22} + e_{23}^{21} + e_{23}^{19} + e_{23}^{15} + e_{23}^{7} + e_{23}^{14} \in \mathbb{Q}(\sqrt{-23})$.

24.1. $(3^6 1^6, 2^{12}, 23^1 1^1)$. Number of trees: **2**.

$a = (1,14,18)(4,7,21)(6,10,9)(8,20,23)(11,17,13)(12,24,16)$
$b = (1,19)(2,24)(3,12)(4,5)(6,8)(7,22)$
$(9,10)(11,14)(13,20)(15,23)(16,18)(17,21)$

Field of moduli: $\mathbb{Q}(\sqrt{-23})$ (H. Monien [**Mon-14**]).
There are, in total, 42 trees with this passport.

24.2. $(3^8, 2^8 1^8, 23^1 1^1)$. Number of trees: **2**.

$a = (1,22,23)(2,24,4)(3,16,8)(5,10,13)$
$(6,14,11)(7,21,19)(9,15,12)(17,18,20)$
$b = (1,5)(2,9)(3,11)(6,13)(7,17)(8,16)(15,20)(18,23)$

Field of moduli: $\mathbb{Q}(\sqrt{-23})$ (Vatuzov [**Vat-19**]).
There are, in total, 429 trees with this passport.

24.3. $(5^4 1^4, 2^8 1^8, 23^1 1^1)$. Number of trees: **2**.

$a = (1,12,4,17,8)(2,22,3,15,5)(7,13,20,14,24)(11,23,18,16,21)$
$b = (1,6)(2,13)(3,19)(5,9)(7,10)(8,21)(12,20)(18,23)$

Field of moduli: we believe that it is the field $\mathbb{Q}(\sqrt{-23})$.
There are, in total, 45 045 trees with this passport.

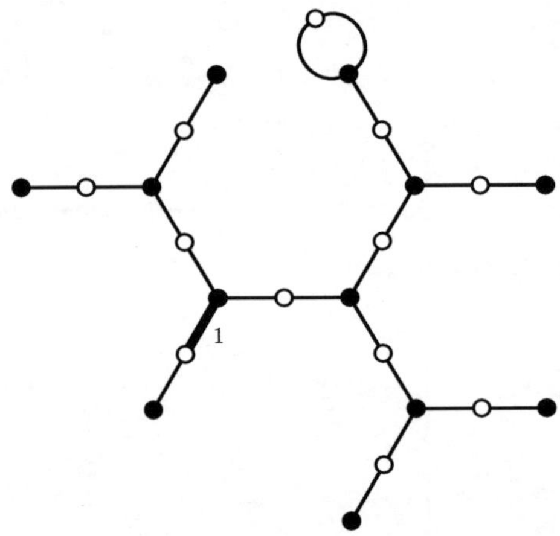

FIGURE 8.33. Group $\mathbf{M}_{24}$: orbit 24.1 of size 2.

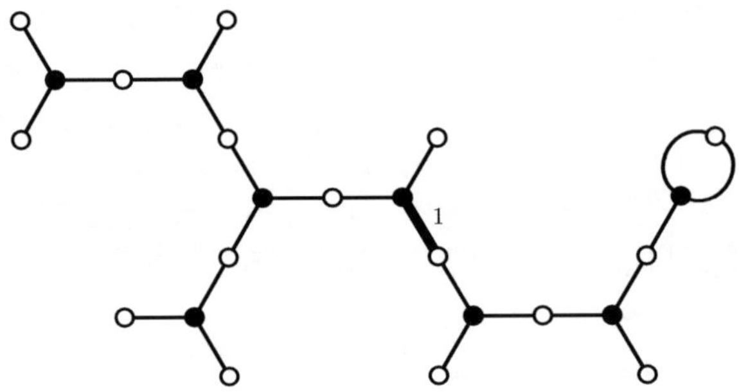

FIGURE 8.34. Group $\mathbf{M}_{24}$: orbit 24.2 of size 2.

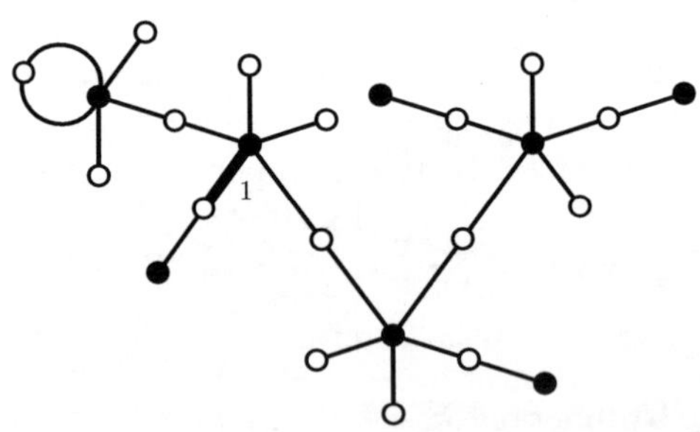

FIGURE 8.35. Group $\mathbf{M}_{24}$: orbit 24.3 of size 2.

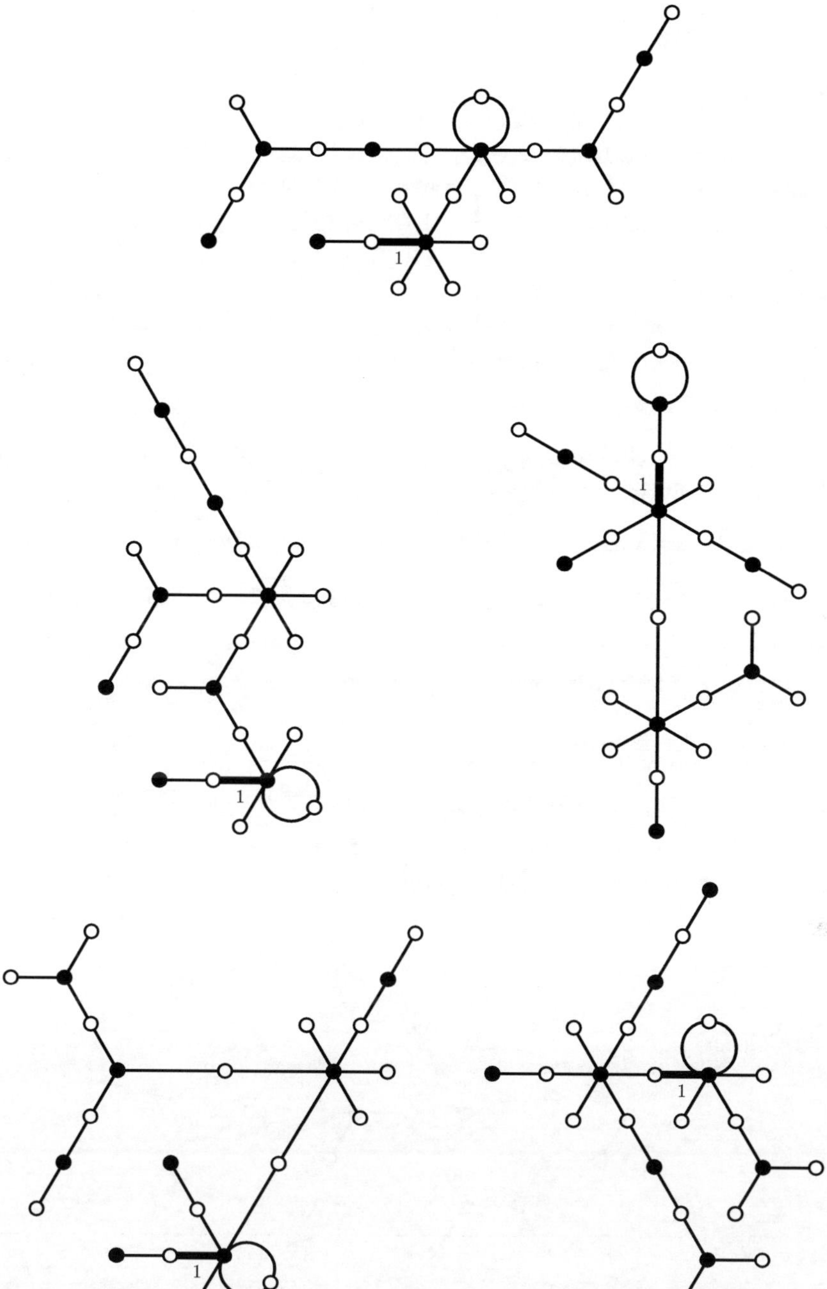

FIGURE 8.36. Group $\mathbf{M}_{24}$: orbit 24.4 of size 10.

24.4. $(6^2 3^2 2^2 1^2, 2^8 1^8, 23^1 1^1)$. Number of trees: **10**.

$a_1 = (1,17,8,14,13,9)(2,19)(3,22,5)(4,21)(6,10,12,23,20,16)(11,18,15)$
$b_1 = (1,7)(2,6)(3,19)(5,24)(10,13)(11,23)(15,21)(16,20)$
$a_2 = (1,17,8,14,13,9)(2,19)(3,22,5)(4,21)(6,10,12,23,20,16)(11,18,15)$
$b_2 = (1,7)(2,21)(3,16)(5,24)(6,15)(8,14)(9,18)(19,20)$
$a_3 = (1,17,8,14,13,9)(2,19)(3,22,5)(4,21)(6,10,12,23,20,16)(11,18,15)$
$b_3 = (1,22)(2,13)(3,5)(6,14)(7,23)(8,24)(15,16)(17,21)$
$a_4 = (1,17,8,14,13,9)(2,19)(3,22,5)(4,21)(6,10,12,23,20,16)(11,18,15)$
$b_4 = (1,7)(2,16)(3,21)(5,18)(8,14)(9,24)(10,22)(12,13)$
$a_5 = (1,17,8,14,13,9)(2,19)(3,22,5)(4,21)(6,10,12,23,20,16)(11,18,15)$
$b_5 = (1,16)(2,24)(4,20)(6,19)(7,12)(8,18)(9,13)(21,22)$

Field of moduli: the splitting field of the polynomial

$$P = a^{10} - 3a^9 + 8a^8 - 16a^7 + 36a^6 - 52a^5 + 92a^4 - 96a^3 + 147a^2 - 89a + 36$$

(Voight [**Voi-16**]). This field is an extension of of the field $\mathbb{Q}(\sqrt{-23})$. Indeed, the above polynomial can be factorized as follows:

$$\begin{aligned} P &= \frac{1}{4} \cdot \left(2a^5 - (3-\sqrt{-23})a^4 - 16a^2 + (12 - 2\sqrt{-23})a - (11 - \sqrt{-23})\right) \\ &\quad \times \left(2a^5 - (3+\sqrt{-23})a^4 - 16a^2 + (12 + 2\sqrt{-23})a - (11 + \sqrt{-23})\right). \end{aligned}$$

The Galois group of P is the wreath product $S_5 \wr S_2$ of order $120^2 \cdot 2 = 28\,800$.

There are, in total, 1 351 350 trees with this passport.

24.5. $(3^6 1^6, 3^6 1^6, 23^1 1^1)$. Number of trees: **2**.

$a = (1,14,18)(4,7,21)(6,10,9)(8,20,23)(11,17,13)(12,24,16)$
$b = (1,13,6)(2,23,5)(3,4,8)(10,12,16)(15,17,19)(20,24,22)$

Field of moduli: we believe that it is the field $\mathbb{Q}(\sqrt{-23})$.

There are, in total, 37 044 trees with this passport.

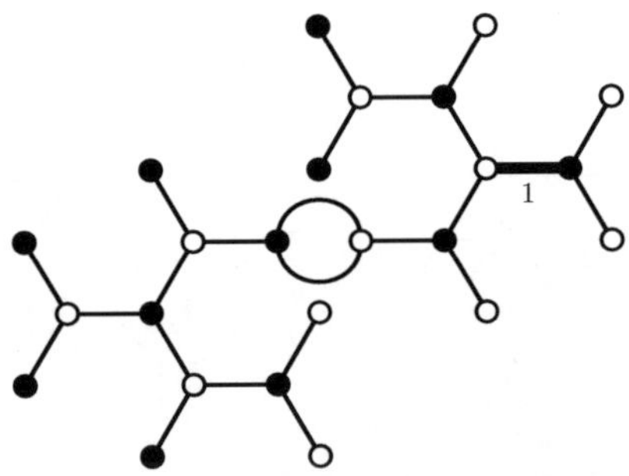

FIGURE 8.37. Group $\mathbf{M}_{24}$: orbit 24.5 of size 2.

$L_5(2)$ of order 9 999 360 PrimitiveGroup(31,10)

Irrationalities in the character table: $e_7^3 + e_7^6 + e_7^5 \in \mathbb{Q}(\sqrt{-7})$;
$e_{15} + e_{15}^2 + e_{15}^4 + e_{15}^8 \in \mathbb{Q}(\sqrt{-15})$;
$e_{31}^3 + e_{31}^6 + e_{31}^{12} + e_{31}^{24} + e_{31}^{17}$, $e_{31}^5 + e_{31}^{10} + e_{31}^{20} + e_{31}^9 + e_{31}^{18}$, $e_{31}^{15} + e_{31}^{30} + e_{31}^{29} + e_{31}^{27} + e_{31}^{23}$: these three numbers, together with their complex conjugates, are the roots of the polynomial $a^6 + a^5 + 3a^4 + 11a^3 + 44a^2 + 36a + 32$.

31.1. $(4^4 2^6 1^3, 2^{12} 1^7, 31^1)$. Number of trees: **6.**

$a_1 = (1,19)(3,8)(4,9,10,16)(6,27)(7,24,22,15)(11,18)$
$(12,23)(13,17,14,30)(20,29)(21,26,25,28)$
$b_1 = (2,24)(5,11)(7,12)(8,23)(10,29)(13,19)(14,15)$
$(16,18)(17,27)(20,28)(22,26)(25,31)$
$a_2 = (1,19)(3,8)(4,9,10,16)(6,27)(7,24,22,15)(11,18)$
$(12,23)(13,17,14,30)(20,29)(21,26,25,28)$
$b_2 = (2,13)(5,15)(7,20)(8,26)(10,17)(11,14)(12,28)$
$(16,25)(18,31)(19,24)(22,23)(27,29)$
$a_3 = (1,19)(3,8)(4,9,10,16)(6,27)(7,24,22,15)(11,18)$
$(12,23)(13,17,14,30)(20,29)(21,26,25,28)$
$b_3 = (2,24)(5,14)(7,12)(8,22)(10,17)(11,15)(13,19)$
$(16,25)(18,31)(20,28)(23,26)(27,29)$

Field of moduli: the splitting field of the polynomial

$$P = a^6 + a^5 + 3a^4 + 11a^3 + 44a^2 + 36a + 32$$

(see [**CaCo-99**]). This field is a cubic extension of the field $\mathbb{Q}(\sqrt{-31})$. Indeed, the above polynomial can be factorized as follows:

$$\begin{aligned} P &= \frac{1}{4} \cdot (2a^3 + (1+\sqrt{-31})a^2 - (5-\sqrt{-31})a - 2(1-\sqrt{-31})) \\ &\quad \times (2a^3 + (1-\sqrt{-31})a^2 - (5+\sqrt{-31})a - 2(1+\sqrt{-31})). \end{aligned}$$

The Galois group of this field is the cyclic group C_6 of order 6.

There are, in total, 12 252 240 trees with this passport.

$ASL_5(2)$ of order 319 979 520 PrimitiveGroup(32,3)

Irrationalities in the character table: $e_7^3 + e_7^6 + e_7^5 \in \mathbb{Q}(\sqrt{-7})$;
$e_{15} + e_{15}^2 + e_{15}^4 + e_{15}^8 \in \mathbb{Q}(\sqrt{-15})$;
$e_{31} + e_{31}^2 + e_{31}^4 + e_{31}^8 + e_{31}^{16}$, $e_{31}^5 + e_{31}^{10} + e_{31}^{20} + e_{31}^9 + e_{31}^{18}$, $e_{31}^7 + e_{31}^{14} + e_{31}^{28} + e_{31}^{25} + e_{31}^{19}$: these three numbers, together with their complex conjugates, are the roots of the polynomial $a^6 + a^5 + 3a^4 + 11a^3 + 44a^2 + 36a + 32$. (Notice that this polynomial is the same as in the orbit 31.1.)

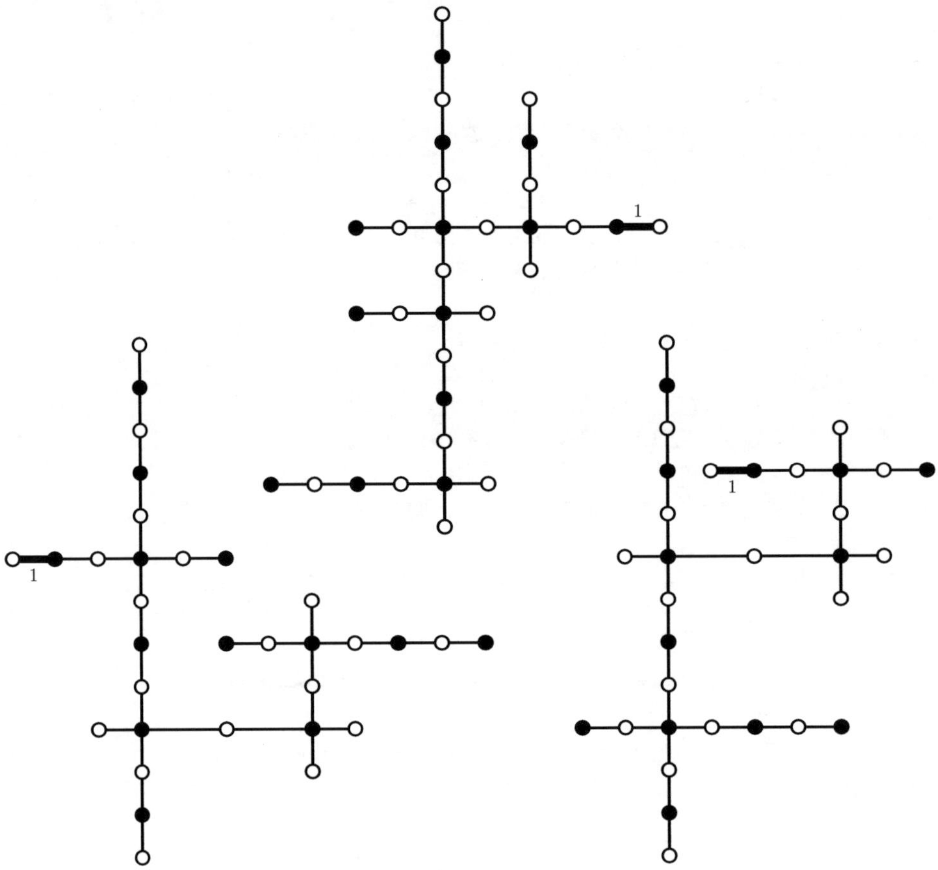

FIGURE 8.38. Group $\mathbf{L}_5(2)$: orbit 31.1 of size 6.

32.1. $(3^{10}1^2, 2^{12}1^8, 31^11^1)$. Number of trees: **6**.

$$a_1 = (2,29,30)(3,24,16)(4,12,19)(5,17,21)(6,13,10)(7,8,28)(9,26,18)(11,15,31)(14,22,25)(20,32,23)$$
$$b_1 = (1,23)(2,24)(7,17)(8,18)(9,25)(10,26)(11,13)(12,14)(15,31)(16,32)(27,29)(28,30)$$
$$a_2 = (2,29,30)(3,24,16)(4,12,19)(5,17,21)(6,13,10)(7,8,28)(9,26,18)(11,15,31)(14,22,25)(20,32,23)$$
$$b_2 = (1,19)(4,18)(6,24)(7,21)(9,15)(10,30)(11,31)(12,14)(13,25)(16,28)(26,32)(27,29)$$
$$a_3 = (2,29,30)(3,24,16)(4,12,19)(5,17,21)(6,13,10)(7,8,28)(9,26,18)(11,15,31)(14,22,25)(20,32,23)$$
$$b_3 = (1,3)(2,16)(4,14)(6,10)(8,12)(13,15)(17,31)(18,20)(19,29)(21,25)(23,27)(30,32)$$

Field of moduli: the splitting field of the polynomial

$$a^6 + a^5 + 3a^4 + 11a^3 + 44a^2 + 36a + 32$$

(Voight [**Voi-16**]). The Galois group of this field is the cyclic group C_6 of order 6.

There are, in total, 218 790 trees with this passport.

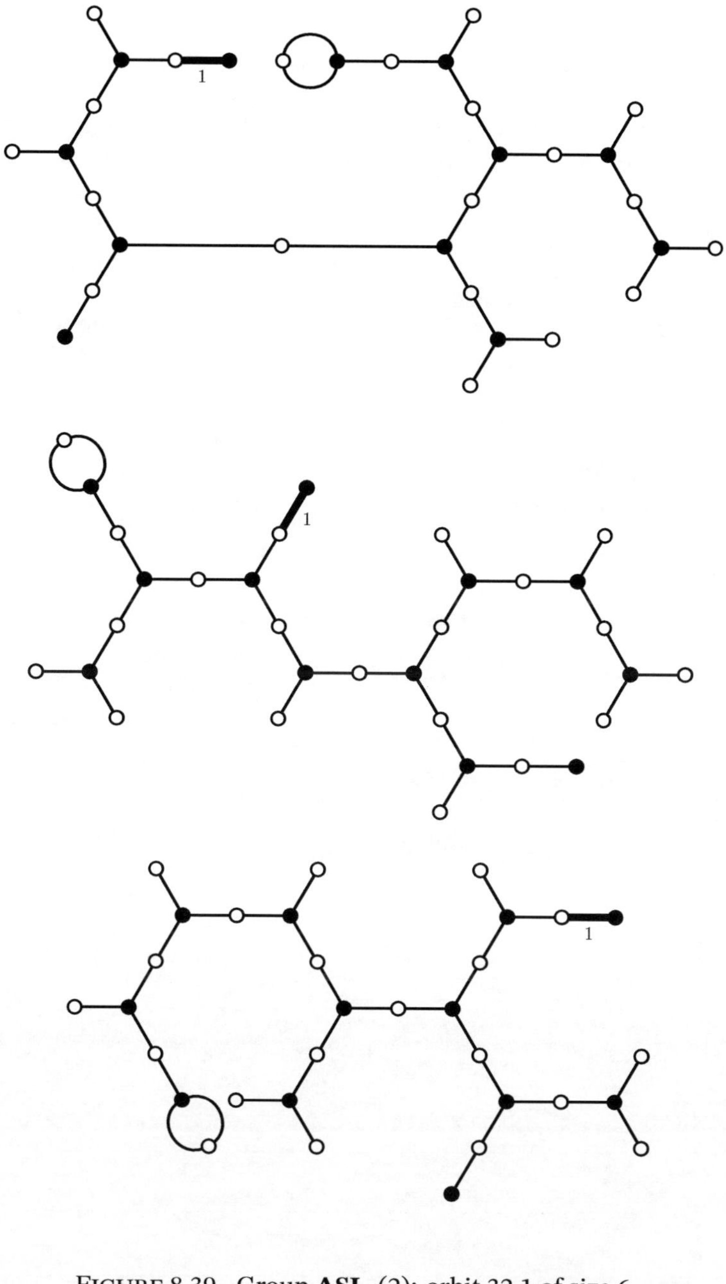

FIGURE 8.39. Group $\mathbf{ASL}_5(2)$: orbit 32.1 of size 6.

CHAPTER 9

A zoo of examples and constructions

The most general (and the most simple) invariant of the Galois action on dessins is the passport. The most sophisticated one is the monodromy group. We have studied both cases in the preceding chapters. In this chapter we mention several other Galois invariants which lead to splitting of various combinatorial orbits into several Galois orbits. Some of them can be expressed in group-theoretic terms but are much easier to detect with the naked eye and do not necessitate complicated calculations with groups.

Let us recall the following definition:

DEFINITION 9.1 (Combinatorial orbit). A *combinatorial orbit* corresponding to a passport is the set of all the trees having this passport.

9.1. Composition

The following proposition is obvious.

PROPOSITION 9.2 (Composition). *Let $f = f(x)$ and $h = h(t)$ be two rational functions such that:*

- *f is a Belyĭ function, with the corresponding dessin D_f;*
- *h is a function (not necessarily a Belyĭ one), all of whose critical values are either vertices or face centers of D_f.*

Then the function $F(t)$ obtained as a composition

$$F(t) = f(h(t)) = f \circ h, \quad \text{that is,} \quad F : \overline{\mathbb{C}} \xrightarrow{h} \overline{\mathbb{C}} \xrightarrow{f} \overline{\mathbb{C}},$$

is a Belyĭ function. If, furthermore, both f and h are defined over $\mathbb{Q}$, then, obviously, the same is true for F. □

The above proposition concerns arbitrary plane dessins, not necessarily trees. Being restricted to weighted trees it gives us the following statement:

COROLLARY 9.3 (Decomposable weighted trees). *Suppose that the functions f and h of the above proposition satisfy the following properties:*

- *the dessin D_f corresponds to a weighted tree, that is, all its finite faces are of degree 1;*
- *h is a* polynomial *all of whose critical values except infinity are vertices of D_f.*

Then all the finite faces of the dessin D_F corresponding to the Belyĭ function $F(t) = f(h(t))$ are of degree 1. If, furthermore, both f and h are defined over $\mathbb{Q}$, then, obviously, the same is true for F.

PROOF. Since h is a polynomial, the only poles of $F = f \circ h$, except infinity, are the preimages of the simple poles of f, i. e., the preimages of the centers of the small faces of D_f. Since h is not ramified over these simple poles, they remain simple for F, and each of them is "repeated" $\deg h$ times. □

EXAMPLE 9.4 (Composition, first example). Consider the following functions:

$$f = -\frac{64\,x^3(x-1)}{8\,x+1}, \qquad u = \frac{1}{5^5}\cdot(t^2+4)^3(3\,t+8)^2.$$

Here f is a Belyĭ function corresponding to the upper left dessin in Figure 9.1, and u is a Belyĭ function corresponding to the lower left dessin.

Substituting $x = u(t)$ in f we obtain a Belyĭ function F corresponding to the dessin shown on the right of Figure 9.1. It is obvious that the *combinatorial* orbit of the dessin D_F consists of more than one element: for example, the petals (both loops and edges) attached to the vertices of degrees 9 and 6 can be cyclically arranged in many different ways. Still, the particular function F we have found belongs to $\mathbb{Q}(t)$ by construction.

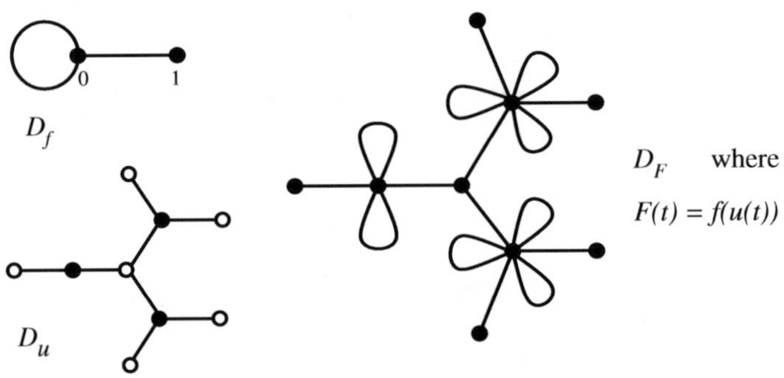

FIGURE 9.1. The dessins D_f and D_F are drawn according to Convention 2.15: only their black vertices are shown explicitly. In the dessin D_u the black vertices are those sent to 0, and the white ones are those sent to 1: both 0 and 1 are vertices of D_f.

Note that the dessins D_f and D_u serving as building blocks for the above example both belong to the classification we have established in Chapter 5: they both correspond to unitrees, and it is their passports that guarantee that they are defined over $\mathbb{Q}$.

A nice property of compositions is that they do not demand difficult computations: the results they produce are, in a way, immediate.

EXAMPLE 9.5 (Composition, second example). Another example, based on the same function f, is as follows. We have

$$f-1 = -\frac{(8\,x^2-4\,x-1)^2}{8\,x+1},$$

so the white vertices of the dessin D_f (which are not shown explicitly in Figure 9.1) are the roots of $8\,x^2-4\,x-1$, that is, they are equal to $(1\pm\sqrt{3})/4$. Now, the critical values of the polynomial

$$v = \frac{1}{3}t^3 - \frac{3}{4}t + \frac{1}{4},$$

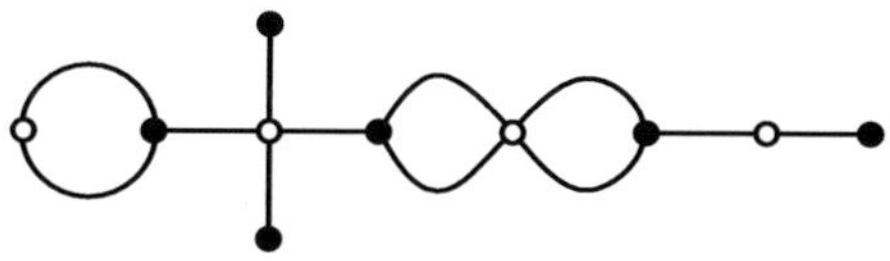

FIGURE 9.2. The dessin D_G corresponding to the function $G(t) = f(v(t))$; this time not all white vertices are of degree 2, therefore we show them explicitly.

that is, the values of v at the roots of $v' = t^2 - 3/4$, are equal to exactly $(1 \pm \sqrt{3})/4$. Therefore, the composition $G(t) = f(v(t))$ is once again a Belyĭ function, and all its poles except infinity are simple. The corresponding dessin is shown in Figure 9.2.

It is obvious that the dessin D_G is not the only one having passport $(3^3 1^3, 4^2 2^2, 9^1 1^3)$. For example, the two "vertical" edges can both be put above, or can both be put below the horizontal axis, or they can be cut off and attached to the leftmost white vertex of degree 2 (the one on the loop), or to the rightmost one (the one on the horizontal segment). However, the dessin of Figure 9.2 is defined over $\mathbb{Q}$ by construction.

Now the dessins D_F and D_G, being defined over $\mathbb{Q}$, may themselves serve for a similar construction: if, for example, w is a polynomial with coefficients in $\mathbb{Q}$ whose critical values are vertices of D_G, then the function $H = G \circ w = f \circ v \circ w$ is a Belyĭ function corresponding to a dessin D_H, all of whose finite faces are of degree 1. In such a composition, only f has to be a Belyĭ function while the subsequent terms (which precede f in the composition) may have more than three critical values.

REMARK 9.6 (Symmetric trees). The group of the orientation preserving automorphisms of a plane tree is always cyclic. If it is C_k then the Belyĭ function for the corresponding map is $F(x) = f(x^k)$ where f is the Belyĭ function for the map corresponding to a single branch of the tree (the vertex of this branch which will become the center of the symmetric tree must be put to the origin). Among the unitrees classified in Chapter 5, the trees N and and R are symmetric (of order 3 and 2 respectively). Some members of the infinite series of unitrees may also be symmetric (for special values of parameters).

We may also recall that the series of unitrees H and I are compositions: see Section 6.7, page 77. Note also that the multiplication of all the weights of the edges of a tree by a factor d can also be represented as a composition with the following Belyĭ function:

$$f(x) = \frac{x^d}{x^d - (x-1)^d}, \qquad \text{hence} \qquad f(x) - 1 = \frac{(x-1)^d}{x^d - (x-1)^d}.$$

REMARK 9.7 (Composition and imprimitive groups). The notion of composition is intimately related to that of the monodromy group, as is shown by Theorem 7.5, page 89: a dessin is a composition of two or more smaller dessins *if and only if* its monodromy group is imprimitive.

This property is valid not only for the weighted trees and even not only for dessins (whatever be their genera) but for arbitrary coverings of topological spaces. We will not dwell on this subject here.

9.2. Difference of powers over $\mathbb{Q}$: infinite series

One of the sources of inspiration for our study was the paper [**BeSt-10**] by Beukers and Stewart. In this paper, the authors consider only the case of powers of polynomials, and only over $\mathbb{Q}$. Namely, they look for polynomials A and B defined over $\mathbb{Q}$ for which the degree $\deg(A^p - B^q)$ attains its minimum. The degrees of polynomials in question are $\deg A = qr$, $\deg B = pr$ where the parameter r is allowed to be greater than one. The passport of the corresponding tree is thus (p^{qr}, q^{pr}).

As is usual in the theory of dessins d'enfants, the authors find several infinite series of DZ-pairs (which they call Davenport pairs), and several sporadic examples. We will return to sporadic examples later in this chapter. As to the infinite series, almost all of them are based on compositions (though there are a few examples in which a quadratic combinatorial orbit splits into two Galois orbits). We will not rewrite their paper but will rather give some general idea of how it works. We will not write down the polynomials but will only draw the corresponding trees.

The trees we need to construct should have the following very simple property: all their black vertices have the same degree p, and all their white vertices have the same degree q. Of course, this condition does not guarantee that the tree in question is defined over $\mathbb{Q}$. The latter will follow from the fact that the trees we compose are defined over $\mathbb{Q}$.

EXAMPLE 9.8 ($A^p - B^{p+1}$). The left tree in Figure 9.3 is defined over $\mathbb{Q}$. Take its vertex of degree 1 and rotate the tree p times around this vertex, so that the center of the tree on the right would acquire the degree p. The resulting star-like tree on the right is obviously defined over $\mathbb{Q}$. Now, all the black vertices of this tree are of degree p while all the white ones are of degree $q = p + 1$.

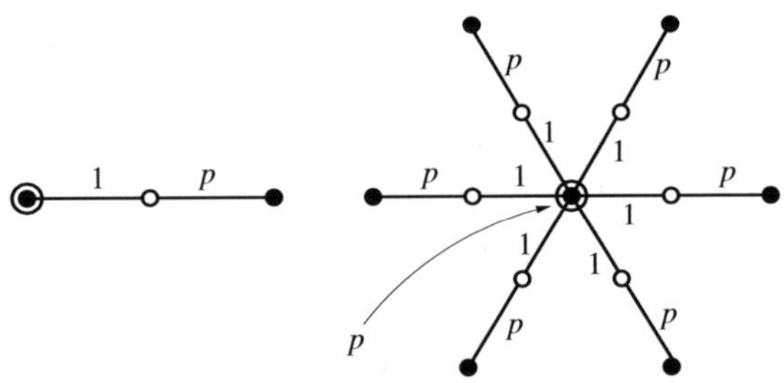

FIGURE 9.3. All the black vertices of the right tree are of degree p, all the white ones are of degree $p + 1$.

EXAMPLE 9.9 (Rotating trees of series E). Consider the tree A in Figure 9.4. It is one of the unitrees of the series E, and therefore it is defined over $\mathbb{Q}$. Its left black vertex (the distinguished one) is of degree 7. This vertex is a bachelor; therefore, we may put it at any rational position, for example, at the origin. The right black vertex is of degree $5 \cdot 5 + 3 = 28$. Therefore, rotating this tree four times around the origin we get five black vertices of degree 28 and 28 white vertices of degree 5 (we leave it to the reader to make a picture). The operation of rotation, in algebraic terms, means inserting x^4 instead of x into the Belyĭ function for the tree A.

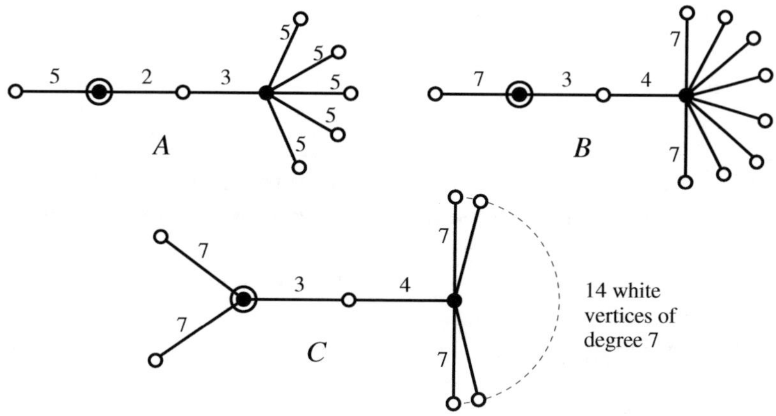

FIGURE 9.4. Examples of trees to be rotated around the distinguished vertex. The number of leaves on the right is 5 for the tree A, 8 for the tree B, and 14 for the tree C. All the leaves of B and C are of degree 7. Rotate A four times around the origin, B six times, C six times.

EXERCISE 9.10 (Generalisation of the above examples). Consider a unitree of the series E and of length 4, with the edges non-leaves having the weights s and t, $s < t$, while the weights of all the leaves are $s+t$. Denote the number of leaves on the left by k, and the number of leaves on the right by l. Find the value of l such that the degree $p = t + l(s+t)$ of the right black vertex is a multiple of the degree $q = s + k(s+t)$ of the left black vertex.

Now we can rotate the tree around the left black vertex p/q times and get a tree with all its white vertices being of the same degree $s+t$, and all the black vertices also being of the same degree p.

For example, for the tree C of Figure 9.4 we have $s = 3$, $t = 4$, $k = 2$, so that $p = 2{\cdot}7+3 = 17$. Now, $l = 14$, so that $q = 14{\cdot}7+4 = 102$. The number 102 is a multiple of 17: $102 = 6 \cdot 17$. Therefore, rotating this tree six times over the distinguished vertex we will get seven black vertices of degree 102. Also, the tree C contains $14 + 3 = 17$ white vertices of degree 7. After being rotated six times it will have 102 of them.

9.3. Polynomials with a relaxed minimum degree condition

Let us return to the problem of the minimum degree of $A^3 - B^2$, the question from which this whole line of research started (see [**BCHS-65**]). Suppose that we are looking for solutions over $\mathbb{Q}$. We have seen that when $\deg A = 2k$, $\deg B = 3k$, we have $\min \deg(A^3 - B^2) = k + 1$. For $k \geq 6$, the computation becomes exceedingly difficult, and anyway there is practically no hope to find a solution over $\mathbb{Q}$. However, if we are not so demanding and accept a solution with the degree of $A^3 - B^2$ slightly greater than $k + 1$, then sometimes we can find a reasonable result.

EXAMPLE 9.11 ($A^3 - B^2$ over $\mathbb{Q}$). Consider the tree shown in Figure 9.5, left. We have here $k = 7$, $\deg A = 14$, $\deg B = 21$. Thus, the minimum degree of $A^3 - B^2$ could in principle be made equal to 8. But here, instead of two vertices of degree 3 we have one vertex of degree 6. One vertex less, hence one face more: the degree of $A^3 - B^2$ is equal to 9 instead of 8. What is so special about this example?

The point is that, even without any computation, we can affirm that this tree is defined over $\mathbb{Q}$. Indeed, the tree on the right is the unitree S considered in Section 6.9.9, hence it is defined over $\mathbb{Q}$. The black vertex of degree 2 of this tree is a bachelor; therefore, we can put it at any rational position without infringing the field. Then, let us put it at $x = 0$. Now, all we have to do is to insert x^3 instead of x into the corresponding polynomials.

Since anyway we have already computed in Chapter 6 the polynomials for the right-hand side tree, we may, without any additional effort, write down the answer. The resulting polynomials look as follows:

$$\begin{aligned} P &= x^6\,(x^{12} + 24x^9 + 176x^6 - 2816)^3, \\ Q &= (x^{21} + 36x^{18} + 480x^{15} + 2304x^{12} - 3840x^9 - 55\,296x^6 \\ &\quad - 14\,336x^3 + 221\,184)^2, \\ R &= 2^{22} \cdot 3^3\,(x^9 + 17x^6 + 56x^3 - 432). \end{aligned}$$

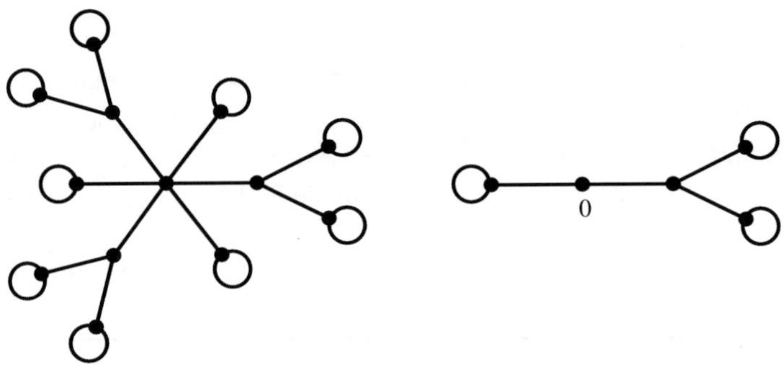

FIGURE 9.5. The map on the left represents two polynomials A and B, of degrees $2k = 14$ and $3k = 21$ respectively, such that $\deg(A^3 - B^2) = 9$. Thus, the degree of the difference does not attain its minimum value $k + 1 = 8$, but in return both A and B are defined over $\mathbb{Q}$.

EXAMPLE 9.12 (Polynomial R with a multiple root). When all the roots of A and B are distinct, the Euler formula ensures that the polynomial R has $k + 1$ distinct roots. Let us accept R with a multiple root, which would mean that one of the inner faces is of degree greater than 1. In this case the degree of R will be greater that $k + 1$. The tree in Figure 9.6 gives such and example. It corresponds to $k = 6$, and $\deg R = 9$ instead of 7.

The polynomials for this tree look as follows:

$$\begin{aligned} P &= (x^3 + 3)^3\,(x^9 + 9x^6 + 27x^3 + 3)^3, \\ Q &= (x^{18} + 18x^{15} + 135x^{12} + 504x^9 + 891x^6 + 486x^3 - 27)^2, \\ R &= 1728x^3\,(x^6 + 9x^3 + 27). \end{aligned}$$

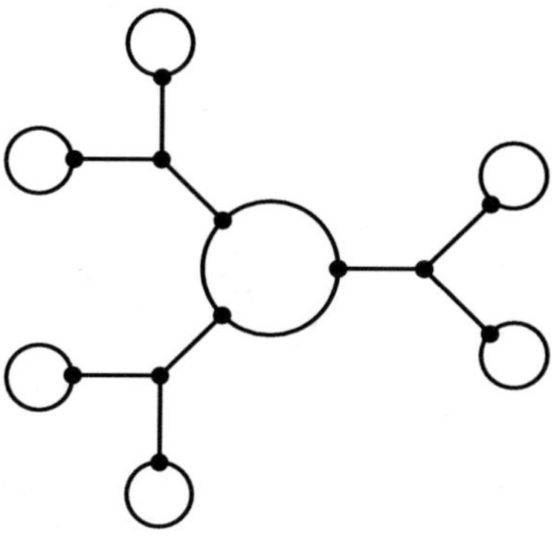

FIGURE 9.6. This map represents two polynomials A and B, of degrees $2k = 12$ and $3k = 18$ respectively, such that the degree $\deg(A^3 - B^2) = 9$. Thus, the degree of the difference does not attain its minimum value $k + 1 = 7$, but in return both A and B are defined over $\mathbb{Q}$.

This line of research is not yet pushed to its logical limits. It would be interesting to construct an infinite series of such examples and to see if it is possible to make the difference between the optimal degree and the degree "over $\mathbb{Q}$" bounded by a constant.

We may, however, mention a result of Dujella [**Duj-10**] which almost achieves this goal, except that the polynomials in question are not coprime. Namely, using as building blocks the polynomials A, B, C defined in (6.9), (6.10) and (6.11) (page 62) with the parameters $s = t = 1$, Dujella constructs an infinite series of pairs of polynomials (S, T) with the following properties:

- $\deg S = 2k$, $\deg T = 3k$;
- S and T are *not* coprime;
- $\deg(S^3 - T^2) = k + 5$, so that the minimum degree $k + 1$ is not attained, though the discrepancy remains bounded;
- in return, *S and T are defined over $\mathbb{Q}$*.

9.4. Duality and self-duality

A dual to a map is usually constructed as follows. First, one puts a new vertex inside every face of the initial map: this vertex is called the "center" of the face. Then, the centers of the adjacent faces are connected by edges in such a way that every edge of the initial map is crossed in its "middle point" by a new edge. A dual of a dual is the initial map.

For the *bicolored maps* a specific variant of the above construction is used, when only black vertices are considered as vertices, while the white vertices play the role of the edge midpoints. (This point of view corresponds to that of *hypermaps* introduced by Cori: see [**Cor-75**], [**Wal-75**].) An association is thus made between the faces of the initial map and the *black vertices* of the dual map while the white vertices belong to both maps. More exactly, a center of a face is connected by edges with all the white vertices lying on the border of the face: see an example in Figure 9.7 where the initial

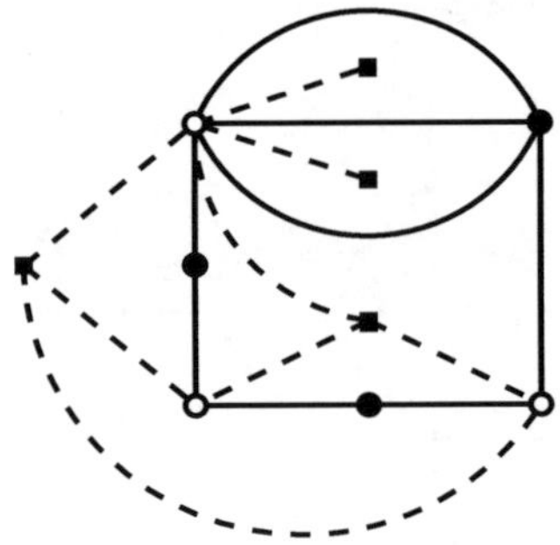

FIGURE 9.7. A bicolored map (solid line), and its dual (dashed line). The black vertices of the dual map are designated by squares. The white vertices belong to both maps.

map is shown by a solid line, and its dual, by a dashed line; the black vertices of the dual map are designated by little squares.

REMARK 9.13 (Black and white). At this point we would like to refer the reader to the discussion of duality on pages 88–89. There are two dualities: black vertices vs. faces and white vertices vs. faces. Here, in order not to repeat everything twice, we consider only the first version of the two.

From the point of view of Belyĭ functions, if $f(x)$ is a Belyĭ function for the initial map, then $1/f(x)$ is a Belyĭ function for its dual. Indeed, $1/y$ interchanges 0 and ∞ while leaving 1 untouched. Therefore, the former poles become roots (i. e., black vertices), and vice versa.

DEFINITION 9.14 (Self-dual map). A bicolored map is called *self-dual* if it is isomorphic to its dual map.

Of course, the fact that a map is self-dual does not mean that $f = 1/f$ where f is its Belyĭ function. It means that

$$1/f(x) = f(w(x)) \tag{9.1}$$

where $w(x)$ a linear fractional transformation of the variable x. The self-duality is an invariant of the Galois action. Indeed, consider the following statement: "There exists a linear fractional transformation $w(x)$ such that (9.1) holds". If this statement is true for a function and is false for another one then these functions cannot belong to the same Galois orbit.

A weighted tree represents a map all of whose faces except one are of degree 1. Therefore, its dual map must have all its black vertices except one being of degree 1. This can only happen if the dual map corresponds to a weighted tree of diameter 4: it has a black vertex of a degree greater than 1 (its central vertex), while all its black leaves are of degree 1. Therefore, if we are interested in self-dual maps which correspond to weighted trees then we must consider only the trees of diameter 4. The condition on the branches of such trees in order for them to be dual to each other is shown in Figure 9.8.

Now we are ready to give an example where the self-duality leads to a splitting of a combinatorial orbit into two Galois orbits.

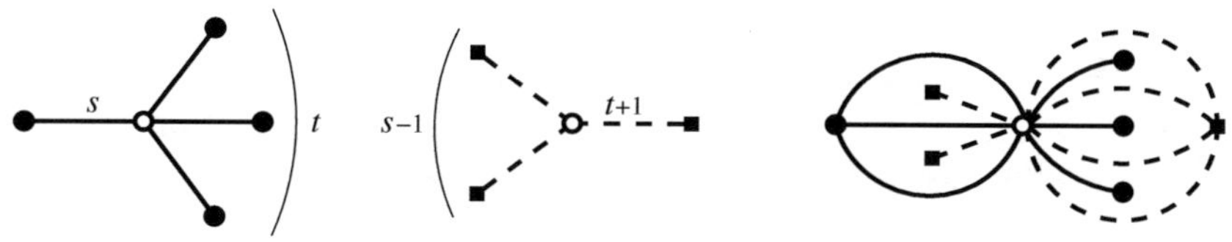

FIGURE 9.8. Two branches of weighted trees of diameter 4 dual to each other. The figure on the right shows how these branches, represented as maps, fit one another.

EXAMPLE 9.15 (Self-duality as a Galois invariant). Let us take two integers p and q, $p < q$, and consider the following passport of degree $n = 2p + 2q - 2$:

- there is a black vertex of degree $p+q$ (the center), and $p+q-2$ black vertices of degree 1 (the leaves);
- there are two white vertices, of degrees $2p-1$ and $2q-1$ respectively;
- the above data imply that the trees have $p+q$ weighted edges (that is, the underlying topological tree has $p+q$ edges), and therefore the outer face is of degree $p+q$, the same as the degree of the central black vertex.

There are $2p-1$ trees with this passport. Their general appearance is shown in Figure 9.9. Here the parameters take the following values: $s = 1, 2, \ldots, 2p-1$ while

$$t = (p+q) - s, \quad k = (2p-1) - s, \quad l = (2q-1) - t.$$

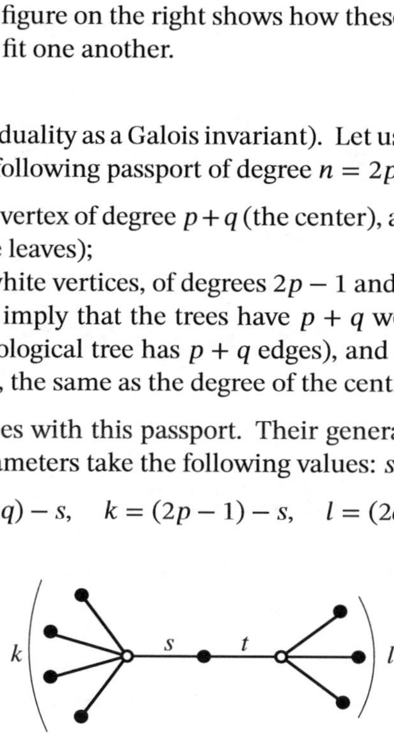

FIGURE 9.9. Here $s+t = p+q$, $s+k = 2p-1$, $t+l = 2q-1$, where $1 < p < q$. The combinatorial orbit consists of $2p-1$ trees but it splits into at least two Galois orbits since exactly one of these trees is self-dual, the one with $s = p$ and $t = q$.

Among all these trees, only one is self-dual: it corresponds to the values $s = p$, $t = q$, $k = p-1$, and $l = q-1$. Therefore, this tree is defined over $\mathbb{Q}$.

(In this example each branch is dual to itself. An attempt to make one branch dual to the other one leads to the equality $p = q$, but we have supposed that $p < q$.)

Let us make a few remarks on the computation of the correponding DZ-pair. Put the white vertices at the points $x = -1$ and $x = 1$ so that

$$Q(x) = (x+1)^{2p-1}(x-1)^{2q-1}$$

(notice that both powers are odd). Observe now that this polynomial is "antipalindromic": if we write it as

$$a_n x^n + a_{n-1}x^{n-1} + \ldots + a_1 x + a_0,$$

then $a_n = -a_0$, $a_{n-1} = -a_1$, ... This fact trivially follows from the equality $x^n \cdot Q(1/x) = -Q(x)$. Because of this, the coefficient in front of the "middle" degree $n/2 = p+q-1$ is zero. Therefore, if we take the higher degrees from $2p+2q-2$ to $p+q$, what will remain is a polynomial of degree $p+q-2$:

$$Q(x) = x^{p+q} \cdot A(x) - R(x),$$

where $\deg A = \deg R = p+q-2$. Setting now $P(x) = x^{p+q} \cdot A(x)$, we see that (P,Q) is a DZ-pair with the required properties. Notice that the polynomial $R(x)$ is reciprocal to $A(x)$. Geometrically, this means that if $x_1, x_2, \ldots, x_m$ are the positions of the black vertices of degree 1 (here $m = p+q-2$), then the centers of the faces of degree 1 are $1/x_1, 1/x_2, \ldots, 1/x_m$. Together with the fact that the position of the black vertex of degree $p+q$ is $x=0$ while the center of the face of degree $p+q$ is ∞, this shows that the map in question is indeed self-dual.

EXAMPLE 9.16. Let us take, for example, $p=2, q=5$. Then

$$\begin{aligned} Q(x) = (x+1)^3(x-1)^9 &= x^{12} - 6x^{11} + 12x^{10} - 2x^9 - 27x^8 + 36x^7 \\ &\quad - (1 - 6x + 12x^2 - 2x^3 - 27x^4 + 36x^5) \\ &= x^7 \cdot A(x) - R(x) = P(x) - R(x), \end{aligned}$$

where $\deg A = \deg R = 5$ and $R = A^*$.

EXERCISE 9.17 (All trees are self-dual). Draw all the five trees with the passport $(6^1 1^2, 3^2 1^2, 6^1 1^2)$ and verify that they are all self-dual. Therefore, for them the self-duality cannot serve as a Galois invariant leading to a splitting of the combinatorial orbit. However, this combinatorial orbit does split into three Galois orbits since one of these trees is symmetric and yet another one has a special monodromy group (see the orbit 8.14).

9.5. A "historic" sporadic example

The enumerative combinatorics works with huge sets of objects which it does not have an intention to distinguish among themselves; the goal is just to count them. The theory of dessins d'enfants is situated at the opposite extreme: its interest is concentrated on specific examples. It would be appropriate to recall the Littlewood's phrase about Ramanujan cited by Hardy: "Every positive integer was one of his personal friends". The same is true with us: we try to make friends with as many particular dessins as possible.

The combinatorial orbit considered in this section is shown in Figure 9.10. It is markworthy for the reason that different authors returned to it, or to some of its elements, many times over the period of 40 years. The Belyĭ function for a was computed by Birch and already appeared in [**BCHS-65**]. The one for d was computed 35 years later by Elkies [**Elk-00**]. All the four Belyĭ functions were independently computed by Shioda [**Shi-05**]. In particular, he found out that the orbit $\{b,c\}$ is defined over the field $\mathbb{Q}(\sqrt{-3})$. Shioda already used as a starting point the four dessins while the other authors apparently made a "blind" search.

So, consider the set of dessins shown in Figure 9.10. They constitute a combinatorial orbit for the passport (α,β,γ) where $\alpha = 3^{10}$, $\beta = 2^{15}$, and $\gamma = 24^1 1^6$. As we have already told, this combinatorial orbit splits into three Galois orbits.

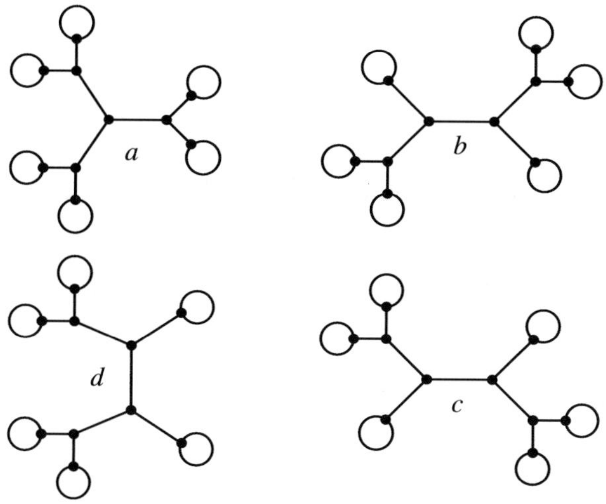

FIGURE 9.10. This combinatorial orbit, corresponding to the passport (α, β, γ) where $\alpha = 3^{10}$, $\beta = 2^{15}$, $\gamma = 24^1 1^6$, splits into three Galois orbits: $\{a\}$, $\{b, c\}$, and $\{d\}$. The dessins a and d are defined over $\mathbb{Q}$.

Namely: the dessin a is the only one having a rotational symmetry of order 3 around a black vertex. Therefore, the singleton $\{a\}$ constitutes a Galois orbit. Two dessins b and c are the only ones which have rotational symmetry of order 2, the center being a white vertex (recall that the white vertices, being all of degree 2, are not shown explicitly in the picture). Therefore, the set $\{b, c\}$ must also be taken apart from the combinatorial orbit. *A priori*, there are two possibilities: b and c may make two Galois orbits, both defined over $\mathbb{Q}$, or they may make a single Galois orbit defined over a quadratic field. But any map whose Belyĭ function is defined over a real field must be axially symmetric since it remains invariant under the complex conjugation. This observation excludes the possibility of two orbits over $\mathbb{Q}$, and it also excludes a real quadratic field. We may conclude that the set $\{b, c\}$ constitutes a single Galois orbit defined over an imaginary quadratic field.

The dessin d is not symmetric and does not have any other specific combinatorial features (the axial symmetry is not a Galois invariant). But it remains solitary, and therefore it constitutes a Galois orbit all by itself. Since the orbits $\{a\}$ and $\{d\}$ both consist of a single element, their Belyĭ functions are defined over $\mathbb{Q}$. Thus, the dessin d is defined over $\mathbb{Q}$ for no other reason than the fact that *it remains alone after all the other Galois orbits being taken away.*

Our combinatorial approach does not make the computational part of the work any easier. Its advantage is elsewhere. It consists in the fact that, before any computation, we may be sure of the following:

- There exist exactly four non-equivalent Belyĭ functions with the passport $(3^{10}, 2^{15}, 24^1 1^6)$; here "non-equivalent" means that they cannot be obtained from one another by a linear fractional change of variables.
- Belyĭ functions corresponding to a and d are defined over $\mathbb{Q}$.

- Belyĭ function corresponding to a is a rational function in x^3 (because of the threefold symmetry of the dessin a).
- Belyĭ functions for the orbit $\{b, c\}$ are defined over an imaginary quadratic field.

The DZ-pair corresponding to d is given at the very beginning of our book: see Example 1.2 on page 1. The DZ-pair for a is as follows:

$$\begin{aligned} P_a(x) &= x^3(x^9 + 12x^6 + 60x^3 + 96)^3\,, \\ Q_a(x) &= (x^{15} + 18x^{12} + 144x^9 + 576x^6 + 1080x^3 + 432)^2\,, \\ R_a(x) &= -1728\,(3x^6 + 28x^3 + 108)\,. \end{aligned}$$

Recall that if (P, Q) is a DZ-pair then the corresponding Belyĭ function is $f = P/R$ while $f - 1 = Q/R$. Notice also that one branch of the tree a is the sporadic unitree M, see Figure 6.16, page 83. It therefore suffices to substitute x^3 instead of x into the polynomials corresponding to M.

We do not present here the DZ-pairs corresponding to the orbit $\{b, c\}$ since they are too cumbersome.

More examples similar to the example of this section are given below.

9.6. Some sporadic examples of Beukers and Stewart

As we have already mentioned at the beginning of Section 9.2, the authors of [**BeSt-10**] have found several infinite series, and also several sporadic examples of pairs of polynomials (A, B) defined over $\mathbb{Q}$ such that $P = A^p$, $Q = B^q$, and $\deg(P - Q)$ attains its minimum. The degrees of the polynomials in question are $\deg A = qr$, $\deg B = pr$ where the parameter r may be greater than 1. The passport of the corresponding tree is (p^{qr}, q^{pr}). Here we consider some of the sporadic examples of [**BeSt-10**].

The first such example, for which $(p, q, r) = (5, 2, 2)$, corresponds to our sporadic tree O, page 84. The next one, $(p, q, r) = (5, 3, 1)$, corresponds to the sporadic tree P, page 84. However, the subsequent examples do not correspond to anything we have seen so far. What's the matter?

It turns out that here we encounter the same phenomenon we have already seen in the previous section, namely, the elements of a combinatorial orbit having different orders of symmetry.

EXAMPLE 9.18 (Parameters $(p, q, r) = (7, 3, 1)$). There exist two trees corresponding to the passport $(7^3, 3^7)$: they are shown in Figure 9.11. We see that one of the trees is symmetric, with the symmetry of order 3, while the other one is not. Therefore, this combinatorial orbit splits into two Galois orbits, and hence both trees are defined over $\mathbb{Q}$. The left-hand one corresponds to the example given in [**BeSt-10**]. Note that an axial symmetry is not a Galois invariant.

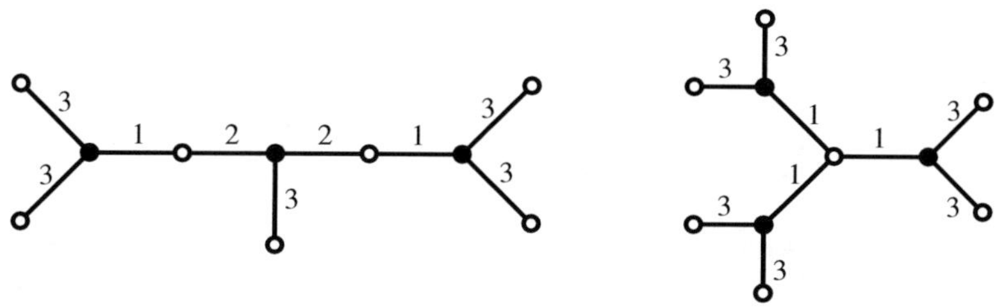

FIGURE 9.11. Two trees corresponding to the passport $(7^3, 3^7)$; one of them is symmetric, the other one is not.

The polynomials corresponding to the asymmetric tree are as follows:

$$\begin{aligned} P &= (x^3 + 18x + 18)^7, \\ Q &= (x^7 + 42x^5 + 42x^4 + 504x^3 + 1008x^2 + 1512x + 3024)^3 \\ R &= 2^4\, 3^3\, (77x^{12} + 5922x^{10} + 6237x^9 + 172\,368x^8 + 366\,606x^7 + 2\,451\,330x^6 \\ &\quad + 7\,314\,300x^5 + 19\,105\,632x^4 + 53\,867\,268x^3 + 82\,260\,360x^2 \\ &\quad + 86\,097\,816x + 62\,594\,856). \end{aligned}$$

EXAMPLE 9.19 ($(p,q,r) = (8,3,1)$ and $(10,3,1)$). The situation for the passports $(8^3, 3^8)$ and $(10^3, 3^{10})$ is similar to the previous one. For the first passport there are two trees, and one of them is symmetric while the other one is not (see Figure 9.12); therefore, both are defined over $\mathbb{Q}$. For the second passport there are three trees (see Figure 9.13). One of them is symmetric with the symmetry of order 2; one is symmetric with the symmetry of order 3; and one is asymmetric. Therefore, all the three trees are defined over $\mathbb{Q}$. In both cases "sporadic" polynomials given in [**BeSt-10**] correspond to the asymmetric trees.

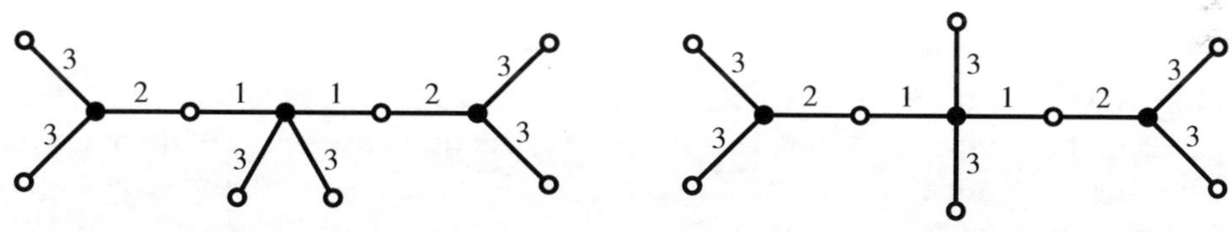

FIGURE 9.12. Two trees corresponding to the passport $(8^3, 3^8)$.

The polynomials for the asymmetric tree with the passport $(8^3, 3^8)$ look as follows:

$$\begin{aligned} P &= (x^3 + 27x + 81)^8, \\ Q &= (x^8 + 72x^6 + 216x^5 + 1620x^4 + 9720x^3 + 24\,300x^2 + 87\,480x + 240\,570)^3, \\ R &= -3^{10}\,(52x^{14} + 6942x^{12} + 21\,816x^{11} + 366\,444x^{10} + 2\,319\,840x^9 \\ &\quad + 13\,129\,047x^8 + 90\,716\,760x^7 + 406\,062\,720x^6 + 1\,812\,830\,544x^5 \\ &\quad + 7\,862\,190\,642x^4 + 23\,694\,237\,936x^3 + 67\,352\,942\,772x^2 \\ &\quad + 173\,534\,618\,376x + 204\,401\,597\,391). \end{aligned}$$

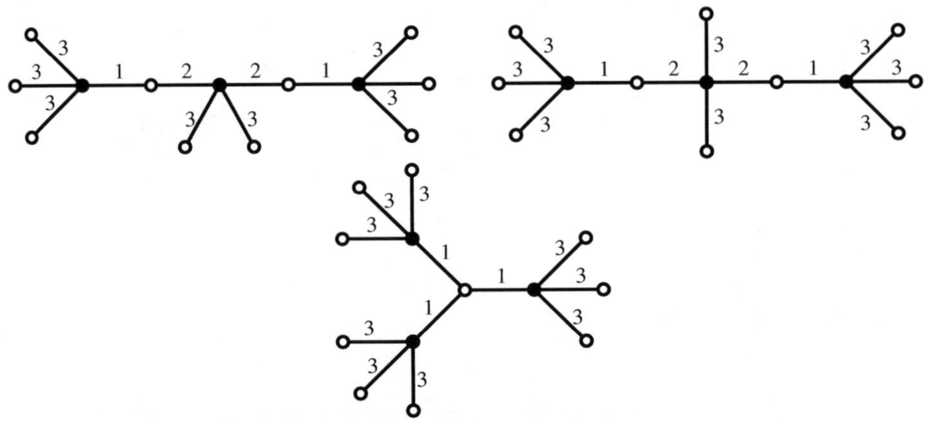

FIGURE 9.13. Three trees corresponding to the passport $(10^3, 3^{10})$.

The polynomials for the asymmetric tree with the passport $(10^3, 3^{10})$ look as follows:

$$\begin{aligned}
P &= (x^3 + 54x + 162)^{10}, \\
Q &= (x^{10} + 180x^8 + 540x^7 + 11\,340x^6 + 68\,040x^5 + 374\,220x^4 \\
&\quad + 2\,449\,440x^3 + 8\,573\,040x^2 + 22\,044\,960x + 57\,316\,896)^3, \\
R &= -2^4\,3^{11}\,(595x^{18} + 201\,960x^{16} + 629\,748x^{15} + 28\,669\,140x^{14} \\
&\quad + 179\,596\,440x^{13} + 2\,460\,946\,860x^{12} + 20\,601\,540\,000x^{11} \\
&\quad + 158\,558\,654\,736x^{10} + 1\,257\,674\,415\,840x^9 + 7\,823\,104\,403\,040x^8 \\
&\quad + 46\,607\,404\,043\,520x^7 + 253\,091\,029\,021\,200x^6 + 1\,120\,772\,437\,834\,752x^5 \\
&\quad + 4\,520\,664\,857\,839\,680x^4 + 15\,435\,507\,254\,345\,280x^3 \\
&\quad + 37\,331\,470\,988\,020\,800x^2 + 62\,014\,139\,393\,904\,000x \\
&\quad + 62\,042\,237\,538\,382\,656).
\end{aligned}$$

EXAMPLE 9.20 (Further sporadic DZ-triples). The next example in **[BeSt-10]** corresponds to the passport $(5^4, 4^5)$. This time, there are three trees: one of them is symmetric with the symmetry of order 2; another one is symmetric with the symmetry of order 4; the third one is asymmetric. All the three are therefore defined over $\mathbb{Q}$.

For the passport $(6^5, 5^6)$ there are four trees. One of them is symmetric with the symmetry of order 5; two are symmetric with the symmetry of order 2; the remaining tree is asymmetric. Therefore, the combinatorial orbit containing four trees splits into three Galois orbits. The asymmetric tree corresponds to the sporadic example given in **[BeSt-10]**.

We leave it to the reader to draw the trees in question.

EXAMPLE 9.21 (When nothing works). At the end of Introduction, we considered a particular tree with the passport $(9^5, 5^9, 13^1 1^{32})$; we show it once again in Figure 9.14. In fact, the combinatorial orbit corresponding to this passport contains 11 trees (see Example 11.17). However, we have to admit that all known combinatorial or group-theoretic invariants fail to explain why this particular tree is defined over $\mathbb{Q}$.

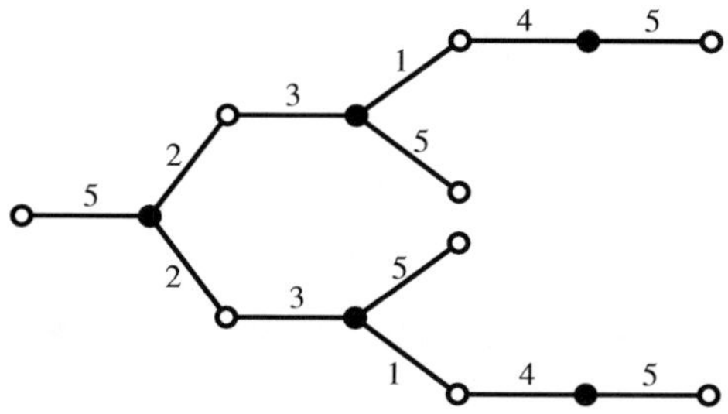

FIGURE 9.14. The combinatorial orbit corresponding to the passport $(9^5, 5^9)$ contains 11 trees. One of them, shown in this figure, is defined over $\mathbb{Q}$. All known combinatorial or group-theoretic invariants of Galois action fail to explain this phenomenon.

We observe that the topological trees corresponding to this passport have 13 edges; thus, the outer face is of degree 13, which is prime. This implies that this tree cannot be a composition. Indeed, the outer face of a tree D_F corresponding to a composition $F = f \circ h$ can only be ramified over the outer face of the tree D_f, so the degree of the outer face of D_F should be equal to the degree of the outer face of D_f multiplied by $\deg h$. But 13 cannot be a product of two integers greater than one.

Now, the monodromy group cannot be special because of Theorem 7.8. The tree cannot be self-dual either since its diameter is greater than 4, etc.

For the moment, this example is the only one of its kind. However, one cannot hope to reduce the whole body of Galois theory to combinatorics. In the next chapter we present a new direction of research: the so-called *Diophantine invariants* of dessins d'enfants. Up to now, this direction remains almost entirely unexplored. But even from this "Diophantine" point of view, the fact that a more or less randomly chosen polynomial has a rational root does not teach us any profound truth. The situation may become interesting if such an example belongs to an infinite series of similar ones and either (a) we can find infinitely many solutions, or (b) we can find several solutions and prove that there are no other ones.

See also the discussion at the end of Section 9.7.

REMARK 9.22 (Without trees). There are no trees in the paper [**BeSt-10**]; all the above sporadic DZ-pairs, as well as several other ones, are found by a brute force computation using Gröbner bases. The above example demonstrates the limitations of our approach. The main message of this book is that it is "almost always" possible to predict "almost all" properties of DZ-polynomials by merely drawing trees. However, we would be unable to predict the existence of an example like 9.21 just by drawing trees.

EXAMPLE 9.23 (An example with many invariants). The combinatorial orbit shown in Figure 9.15 corresponds to the passport $(8^1 1^2, 2^4 1^2)$. It consists of sixteen trees and splits into four Galois orbits.

The first tree is "special": its monodromy group is $\mathrm{PGL}_2(9)$ (see Chapter 8). The next two trees are symmetric. The remaining 13 trees are not symmetric and all have the monodromy group S_{10}. However, the first five of them are self-dual while the other eight are not.

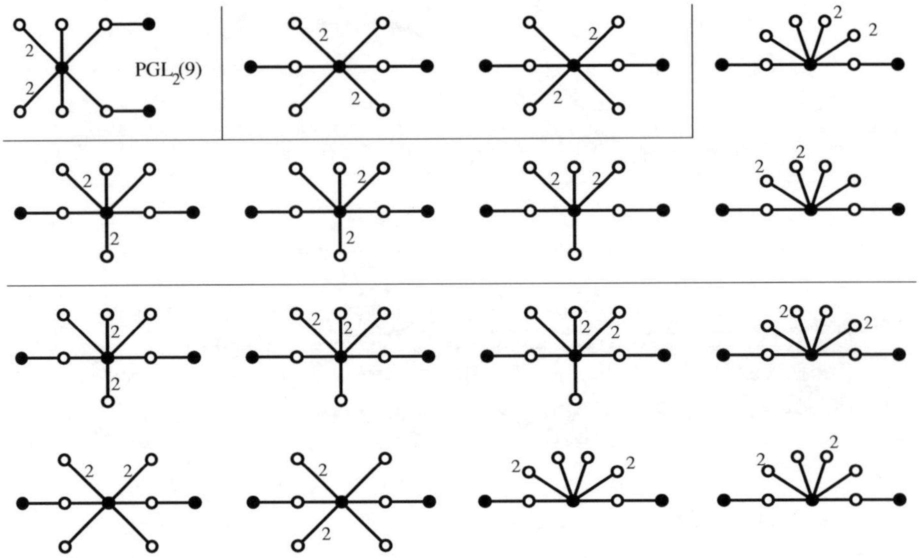

FIGURE 9.15. This combinatorial orbit splits into four Galois orbits.

Notice however that in order to be sure that there are no further accidental splitting one must compute the corresponding Belyĭ functions. Without such a computation one can only say that there are *at least* four orbits. We did the needed computation, so in this particular example we may affirm that there are indeed exactly four orbits.

We must point out that this example is in no way generic. In the absolute majority of cases a combinatorial orbit represents a unique Galois orbit.

9.7. Sporadic examples of "megamap invariant"

The three trees shown in Figure 9.16 are defined over $\mathbb{Q}$.

(A) One of the four trees with the passport $(4^1 2^1 1^1, 3^1 2^2, 5^1 1^2)$.

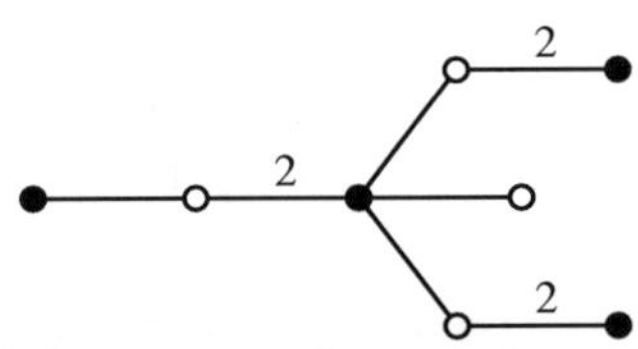

(B) One of the 17 trees with the passport $(5^1 2^2 1^1, 3^3 1^1, 7^1 1^3)$.

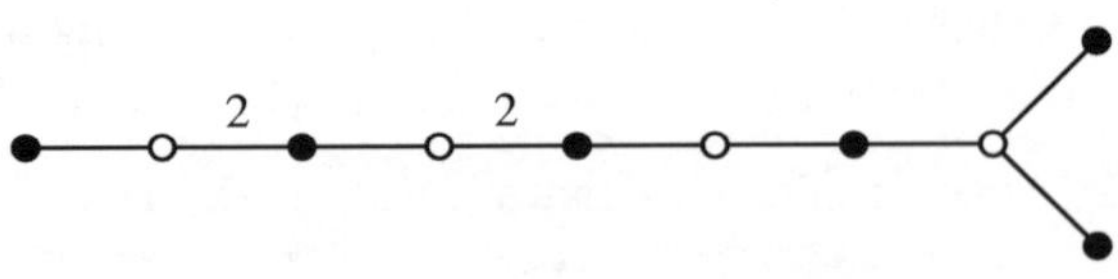

(C) One of the 43 trees with the passport $(3^2 2^1 1^3, 3^3 2^1, 9^1 1^2)$.

FIGURE 9.16. Three weighted trees defined over $\mathbb{Q}$.

This fact is far from being obvious: the corresponding combinatorial orbits contain 4, 17 and 43 trees, respectively, the trees are neither self-dual nor decomposable, and their monodromy groups are not special. Nevertheless the corresponding DZ-polynomials can be written as follows:

$$\begin{aligned} P_7 &= x^4(x+3)^2(x+5), \\ Q_7 &= (x+1)^3(x^2+4x-2)^2, \\ R_7 &= 4(2x+1)(3x-1); \end{aligned}$$

$$\begin{aligned} P_{10} &= x^5(16x^2-8x+21)^2(x+1), \\ Q_{10} &= (4x^3+x^2+3x+3)^3(4x-3), \\ R_{10} &= 27(5x^3+5x+3); \end{aligned}$$

$$\begin{aligned} P_{11} &= 9x^2(3x^2+18x+25)^3(3x^3+6x^2-27x+20), \\ Q_{11} &= (9x^3+54x^2+45x-100)^3(x+1)^2, \\ R_{11} &= 50000(x+4)(2x+5). \end{aligned}$$

Is there another invariant which would explain the splitting of these combinatorial orbits? Yes, there is one, though not easily detectable. We will describe the idea briefly since it goes beyond the scope of this book.

Let us fix four partitions $\lambda_1, \lambda_2, \lambda_3, \lambda_4 \vdash m$ and consider coverings of degree m of the sphere, ramified over *four* points so that the ramification multiplicities over these points are $(\lambda_1, \lambda_2, \lambda_3, \lambda_4)$. The covers are Riemann surfaces of genus g where g is determined by the Riemann–Hurwitz formula. The family of such coverings (if it is not empty) constitutes a one-dimensional stratum in the *Hurwitz space* $H_{g,m}$, which is the space of pairs (X, f), where X is a Riemann surface of genus g and f is a meromorphic function of degree m on X. The family itself is also often called Hurwitz space, but we prefer to call it *Fried family* (see **[Sha-19]**) to avoid misunderstanding.

Every Fried family is endowed with a naturally defined Belyĭ function: the cross-ratio of the four branch points, see **[Fri-77]**, **[DDH-89]**, **[Oga-13]**, **[Oga-16]**. The corresponding dessin d'enfant is called *megamap*, and it can be determined combinatorially using an action of the *Hurwitz braid group* $\mathcal{H}_4$, see details in **[Zvo-01]**, **[LaZv-04]**, **[Rob-15]**. Notice that there exist various versions of this construction, depending on whether the set of partitions $(\lambda_1, \lambda_2, \lambda_3, \lambda_4)$ is ordered or not.

It may happen that the Fried family is a reducible algebraic curve. In this case the corresponding megamap is a disconnected dessin d'enfant. The absolute Galois group $\mathrm{Aut}(\overline{\mathbb{Q}}|\mathbb{Q})$ acts on the connected components of the megamap permuting them. Whenever a component of the megamap can be uniquely distinguished among other components using combinatorial invariants (genus, number of edges, passport, monodromy group, self-duality), we may conclude that this dessin is defined over $\mathbb{Q}$. In particular, this argument always works when the megamap is connected.

One can check that the dessins in Figure 9.16 are the megamaps of the Fried families with the following passports: $(4^1 1^1, 3^1 1^2, 2^2 1^1, 2^1 1^3)$, $(5^3, 3^5, 6^1 1^9, 2^1 1^{13})$ and $(5^1 4^1, 3^3, 2^2 1^5, 2^1 1^7)$ of degrees $m = 5$, 15 and 9 respectively. We therefore conclude that they are defined over $\mathbb{Q}$, and the computation of their Belyĭ functions just confirms it.

The megamap construction is a plentiful source of Belyĭ functions defined over $\mathbb{Q}$. A profusion of examples of this kind were given in the conference talk **[Rob-15]** by David Roberts and in his paper **[Rob-16]**. For example, the tree (A) of Figure 9.16

corresponds to the megamap-tree of Figure 2.1 of [**Rob-16**]. We will present some non-trivial infinite series of megamaps of genus 0 defined over $\mathbb{Q}$ in a forthcoming paper.

The calculation of generating permutations of a megamap is much easier than computing its Belyĭ function. Unfortunately, we don't have an algorithm which would recognize megamaps among other dessins d'enfants. For example, we don't know if the tree of Example 9.21 is a megamap. The three above examples were found by a computer-aided blind search. Namely, we considered a huge number of Fried families in order to find megamaps which would be weighted trees. It is very likely that there are more examples of this kind.

9.8. One more application due to David Roberts

In the paper [**Rob-14**] by Roberts[1], a remarkable application of weighted trees is given. An explicit construction is proposed for polynomials of a very high degree and with a very small set of primes of bad reduction. The most extreme example is given by a polynomial $f \in \mathbb{Z}[x]$ of degree

$$n = 15\,875 = 125 \cdot (128 - 1),$$

with the Galois group of its splitting field equal to S_n, and with the set of primes of the bad reduction equal to $\{2, 5\}$. More exactly, the discriminant of the field is equal to

$$\Delta = -2^{130\,729} 5^{63\,437}.$$

Unfortunately, we cannot present these results here since they would necessitate a voluminous preparatory material.

[1]The year is not indicated in the paper but it is certainly not later than 2014 since Professor Roberts showed this paper to A. Z. in March of 2015.

CHAPTER 10

Diophantine invariants

The theory of dessins d'enfants studies the action of the absolute Galois group $\mathrm{Aut}(\overline{\mathbb{Q}}|\mathbb{Q})$ on bicolored maps, with a particular interest in the search of invariants of this action. In the vast majority of cases, such invariants are of combinatorial and/or group-theoretic nature. Sweet dreams are sometimes expressed in the dessins d'enfants community that it would be desirable to find a *complete* set of such invariants, a sort of "two dessins belong to the same Galois orbit *if and only if* all their invariants are equal". One of the goals of this chapter is to show that such a system of invariants just cannot exist. Namely, there are certain cases when the dessins in question do not present any particular combinatorial or group-theoretic properties, and the Galois splitting is explained by some diophantine relations between certain numerical characteristics of the dessins in question. We have already encountered a similar situation even before any considerations of Galois theory: see Theorem 3.1 (page 19). We call such relations *Diophantine invariants.* Thus, the statement that a complete set of combinatorial or group-theoretic invariants cannot exist is not entirely negative since the Diophantine equations are a remarkable subject in itself. In this chapter we present a particularly beautiful example of this phenomenon, when the question of splitting of a combinatorial orbit of size 2 into two Galois orbits over $\mathbb{Q}$ is reduced to the famous Pell equation.

10.1. Pell's equation: preliminaries

DEFINITION 10.1 (Pell's equation). Let $D > 0$ be an integer which is not a perfect square. Then the Diophantine equation

$$x^2 - Dy^2 = 1 \tag{10.1}$$

is called *Pell's equation.*

The word Diophantine means that we look for solutions in $\mathbb{N}$ or in $\mathbb{Z}$. The name of the British mathematician Pell was erroneously attributed to this equation by Euler: Pell never worked on it. However, to call it otherwise would be incongruous: the readers would not understand what we are talking about.

10.1.1. A brief history of the Pell equation. This innocently looking and inconspicuous equation is a real mathematical jewel. It is studied for more than two thousand years, and people still find something new to say about it. Among the recent publications we may mention a monograph [**JaWi-09**] by Jacobson and Williams (of more than 500 pages!); a problem book [**Bar-03**] by Barbeau; and a scientific-popular brochure [**Bug-10**] by Bugaenko. One of the proofs of the algorithmic undecidability of Hilbert's Tenth problem is based on the properties of Pell's equations: see a short announcement in [**Chu-70**] and a detailed exposition in [**JoMa-91**].

The first name mentioned in relation to the Pell equation is that of Pithagoras (VIth century before n. e.). The books on the history of mathematics do not say what exactly was his contribution to the subject but we can advance a plausible conjecture: since the equation $x^2-2y^2=0$ does not have a solution in integers then let us try the closest one: $x^2-2y^2=1$.

The next appearance of this equation is in a letter by Archimedes to Eratosthenes (IIIrd century before n. e.) concerning the cattle of the god Helios. The full text of the letter, as well as a relevant discussion, may be found in [**JaWi-09**], pages 19–24. In order to establish the number of bulls of Helios one must write down a system of algebraic equations which is reduced to the Pell equation with $D = 410\,286\,423\,278\,424$. It may well be that the whole story is a pure legend. At that time, the positional system was not yet invented, and without its apparatus it is close to impossible to work with huge numbers. Therefore, it is highly improbable that Archimedes was himself able to solve an equation with such an enormous coefficient. But maybe he proceeded in the opposite direction: from a solution to the equation.

It goes without saying that Diophantus (IIIrd century of n. e.) could not overpass this equation. Then, further developments were made by Indian mathematicians: Brahmagupta (VIIth century), Bhaskara II (XIIth century), Narayana Pandita (XIVth century)… and we return to Europe with the British mathematician Brouncker (XVIIth century).

Then comes the omnipresent platoon of Fermat, Euler, Lagrange, Abel, Dirichlet, and certainly many others.

Bhaskara II proposed an algorithm to compute a solution of the Pell equation; this algorithm is still in use today (see, for example, an online Pell's solver [**PellSol**]). Lagrange proved (1768) that this algorithm always finds a solution. As we will see below, this implies that there is always infinitely many solutions.

Abel studied the problem in which x, y and D are polynomials in one variable. For example, for $D = t^2-1$ one has $x_n = T_n(t)$ and $y_n = U_{n-1}(t)$ where T and U are Chebyshev polynomials of the first and second kind.

Starting from the late XVIIIth century, mathematicians studied the ring

$$\mathbb{Z}(\sqrt{D}) = \{x + y\sqrt{D} \mid x, y \in \mathbb{Z}\}$$

and were surprised to discover that this ring possessed infinitely many divisors of unity. Indeed, if (x, y) is a solution of (10.1) then both $x+y\sqrt{D}$ and $x-y\sqrt{D}$ are divisors of unity since their product is equal to one.

And so on…

10.1.2. How to solve Pell's equation. It is known that the Pell equation has infinitely many solutions. All solutions lie on the quarter of the hyperbola

$$\{(x, y) \in \mathbb{R}^2_+ \mid x^2 - Dy^2 = 1\};$$

therefore, it is possible to order them from left to right. There always exists the trivial solution $(x_0, y_0) = (1, 0)$; the next one, the solution (x_1, y_1), is called the *fundamental solution*.

PROPOSITION 10.2 (Solutions of Pell's equation). *Consider the matrix*

$$A = \begin{pmatrix} x_1 & Dy_1 \\ y_1 & x_1 \end{pmatrix} \tag{10.2}$$

where (x_1, y_1) *is the fundamental solution of* (10.1). *Then all the solutions of* (10.1) *are given by the formula*

$$\begin{pmatrix} x_n \\ y_n \end{pmatrix} = A^n \begin{pmatrix} x_0 \\ y_0 \end{pmatrix}.$$

PROOF. If (x_n, y_n) is a solution then the multiplication by the matrix A gives another solution: this fact is established by a trivial verification. Why there are no other solutions? Notice that $\det A = 1$; therefore, the entries of the inverse matrix A^{-1} are integers. Suppose that there is a solution which lies between the nth and $(n+1)$st solutions obtained by the above formulas. Multiplying it by A^{-1} we get a solution lying between $(n-1)$st and nth ones. Repeating this procedure we will finally find a solution between (x_0, y_0) and (x_1, y_1). But the existence of such a solution contradicts the definition of the fundamental solution: (x_1, y_1) is *the first* solution to appear after (x_0, y_0). □

We may also notice that if (x, y) is a solution then

$$(x + y\sqrt{D})(x - y\sqrt{D}) = 1,$$

hence

$$(x + y\sqrt{D})^n(x - y\sqrt{D})^n = 1.$$

Writing down $(x + y\sqrt{D})^n$ as $x_n + y_n\sqrt{D}$ we find infinitely many solutions (x_n, y_n).

We see that the most important step in solving Pell's equation is to find the fundamental solution. Lagrange was the first to prove its existence. A great difficulty consists in the fact that even for moderate values of D the fundamental solution may be very large. For example, for $D = 1021$ the smallest solution after $(x_0, y_0) = (1, 0)$ is

$$\begin{aligned} x_1 &= 198\,723\,867\,690\,977\,573\,219\,668\,252\,231\,077\,415\,636\,351\,801\,801, \\ y_1 &= 6\,219\,237\,759\,214\,762\,827\,187\,409\,503\,019\,432\,615\,976\,684\,540. \end{aligned}$$

Another example: for $D = 410\,286\,423\,278\,424$ the fundamental solution contains 206 545 decimal digits (this information is taken from [**JaWi-09**]). It is clear that no naive algorithms could cope with such examples. However, there do exist many sophisticated and efficient algorithms to solve this problem. For example, the solver [**PellSol**] takes less than one minute to find the above solution for $D = 1021$.

10.1.3. Pell-like equations. Pell-*like* equation is the equation

$$x^2 - Dy^2 = k, \quad k \neq 1.$$

This equation may have either no solutions at all, or infinitely many of them.

EXAMPLE 10.3 (No solutions). Let us take the equation $x^2 - 7y^2 = 3$. Reducing it modulo 7 we get the equation $x^2 = 3 \bmod 7$. But 3 is not a quadratic residue in $\mathbb{Z}_7$. Therefore, the above equation has no solutions.

Consider now the equation $x^2 - 7y^2 = 11$. This time 11 is equal to 4 modulo 7 and thus it is a quadratic residue. Still, this equation has no solutions either. It suffices to examine all the pairs x, y modulo 11.

EXAMPLE 10.4 (Infinitely many solutions, but...). Let us take now the equation $x^2 - 7y^2 = 18$. Its smallest solution is $(x_1, y_1) = (5, 1)$. We obtain infinitely many solutions by the following multiplication:

$$A^n \cdot \begin{pmatrix} 5 \\ 1 \end{pmatrix} \quad \text{where} \quad A = \begin{pmatrix} 8 & 21 \\ 3 & 8 \end{pmatrix}.$$

Here the matrix A is obtained from the fundamental solution $(8, 3)$ of the standard Pell equation $x^2 - 7y^2 = 1$. The problem is that in this way we do not get *all* the solutions. Indeed, after making the first multiplication we jump immediately to the fourth one $(x_4, y_4) = (61, 23)$, missing two intermediate solutions: $(x_2, y_2) = (9, 3)$ and $(x_3, y_3) = (19, 7)$. The set of all solutions is a union of the three infinite series obtained from (x_1, y_1), (x_2, y_2) and (x_3, y_3) by multiplying them by the matrix A.

The algorithm verifying if a Pell-like equation has a solution is rather complicated: see Section 16.3 of the book [**JaWi-09**].

We are lucky: in the example that follows the equation is $x^2 - 2y^2 = 1$, and its fundamental solutionis $(x_1, y_1) = (3, 2)$: indeed, $3^2 - 2 \cdot 2^2 = 1$.

10.2. When a quadratic orbit splits into two rational ones

10.2.1. Combinatorial orbit. We consider the dessins with the following characteristics (see Figure 10.1): the black partition is $\alpha = m^3$, that is, there are three black vertices, each of them of degree m; the white partition is $\beta = 5^1 1^{3m-5}$, that is, there is one white vertex of degree 5 (the "center"), while all the other white vertices are of degree 1; the face partition is $\gamma = (3m-2)^1 1^2$, that is, there is an outer face of degree $3m-2$ and two faces of degree 1.

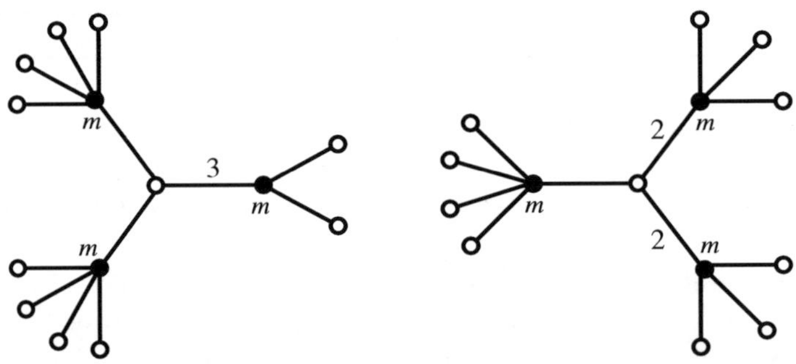

FIGURE 10.1. A combinatorial obit consisting of two trees: black vertex degrees are equal to $m \geq 3$ (in the figure, $m = 5$).

There are two dessins with the above passport. They look as is shown in Figure 10.1. By merely looking at the dessins we see that both trees are defined over a real field. They may constitute a single orbit defined over a real quadratic field, or two separate orbits both defined over $\mathbb{Q}$.

Combinatorially, the trees don't have any particular features which would permit to distinguish them and to put them in separate Galois orbits. From the group-theoretic point of view, there is nothing to say either, as the following proposition shows.

PROPOSITION 10.5 (Monodromy groups for the dessins of Figure 10.1). *For* $m \geq 3$, *the monodromy groups of both dessins of* Figure 10.1 *are* S_{3m} *if* m *is even, and* A_{3m} *if* m *is odd.*

PROOF. Let us first prove that the groups in question are primitive. According to Theorem 7.5 (page 89), we must prove that the corresponding Belyĭ functions are indecomposable.

Let F be a Belyĭ function, and let M_F be the corresponding dessin. Suppose that F is decomposable, that is, $F = g \circ f$ where $\deg(f) > 1$, $\deg(g) > 1$. Here g must be a Belyĭ function while f is not necessarily Belyĭ but its critical values must be either vertices or face centers of the map M_g corresponding to g.

Let A be a face of M_g and $\deg(A) = k$. Then $f^{-1}(A)$ is a set of faces of M_F whose degrees are multiples of k, and the sum of these degrees is equal to $k \cdot \deg(f)$. In our case, both dessins of Figure 10.1 have two faces of degree 1. Therefore, $\deg(f) = 2$ and f (and hence $F = g \circ f$) must be invariant under a symmetry of order 2. But our dessins are not symmetric. Hence, the function F cannot be a composition, and the monodromy groups of both dessins are primitive.

What remains is to apply the classical Jordan's "symmetric group theorem" first appeared in [**Jor-1870**]: it states that a primitive permutation group of degree n which contains a cycle of a prime order $p < n - 2$ is either S_n or A_n. In our example, the monodromy group is of degree $n = 3m$, and it contains a cycle of order 5 (the permutation corresponding to the white vertices). Thus, for $m \geq 3$ it satisfies the conditions of Jordan's theorem.

By the way, for $m = 2$ there exists only one tree, the one on the right in Figure 10.1. Therefore, it is necessarily defined over $\mathbb{Q}$. As to the monodromy group, it is in this case not S_6 but $PGL_2(5)$. □

We conclude that the monodromy group is the same for both trees, so it does not permit us to separate them. What comes to the rescue is the Pell equation. It provides us with a *complete* list of splitting combinatorial orbits.

10.2.2. Belyĭ function and the field of moduli. The computation of the Belyĭ function proceeds as follows. We put the center of the outer face to $x = \infty$; the white vertex of degree 5, to $x = 0$; and let the sum of the positions of the centers of two small faces be equal to 1. Then the Belyĭ function takes the following form:

$$f = K \cdot \frac{(x^3 + ax^2 + bx + c)^m}{x^2 - x + d}.$$

Computing f' we get

$$f' = K \cdot \frac{(x^3 + ax^2 + bx + c)^{m-1} \cdot q(x)}{(x^2 - x + d)^2},$$

where $q(x)$ is a polynomial of degree 4. What remains is to make $q(x)$ proportional to x^4, that is, to equate all the coefficients of $q(x)$, except the leading one, to zero. This gives us four equations for the unknowns a, b, c, d. The factor K is then determined by the condition $f(0) = 1$.

As a result of the computation we find out that all the coefficients of Belyĭ function belong to the real quadratic field $\mathbb{Q}(\sqrt{\Delta})$, where

$$\Delta = 3(2m - 1)(3m - 2). \tag{10.3}$$

Thus, our combinatorial orbit splits into two Galois orbits when, and only when the parameter Δ in (10.3) is a perfect square.

10.3. When the discriminant is a perfect square

Two remarks are in order. First, the numbers $2m-1$ and $3m-2$ are coprime, which can be verified by a direct application of Euclid's algorithm. Second, $3m-2$ cannot be divisible by 3; only $2m-1$ can. We conclude that, in order to make Δ a perfect square, its two factors $3(2m-1) = 6m-3$ and $3m-2$ should both be made perfect squares. Then, writing down

$$6m - 3 = a^2, \qquad 3m - 2 = b^2, \tag{10.4}$$

we observe that

$$a^2 - 2b^2 = 1, \tag{10.5}$$

that is, the pair (a, b) must be a solution of the Pell equation with $D = 2$.

The equalities (10.4) imply that the parameter m is found as

$$m = \frac{a^2+3}{6} \qquad \text{and} \qquad m = \frac{b^2+2}{3}.$$

Therefore, in order to fit into our scheme, the parameter a must be odd and divisible by 3 while b should not be divisible by 3.

It is easy to verify that every other solution of the Pell equation satisfies both conditions. Indeed, the main recurrence

$$\begin{pmatrix} x_{n+1} \\ y_{n+1} \end{pmatrix} = \begin{pmatrix} 3 & 4 \\ 2 & 3 \end{pmatrix} \begin{pmatrix} x_n \\ y_n \end{pmatrix}, \quad \begin{pmatrix} x_0 \\ y_0 \end{pmatrix} = \begin{pmatrix} 1 \\ 0 \end{pmatrix},$$

being taken modulo 3 gives the following sequence:

$$(1,0) \to (0,2) \to (1,0) \to (0,2) \to \ldots$$

The congruence

$$(a, b) \equiv (0, 2) \mod 3$$

means that a is divisible by 3 while b is not. Also, a is always odd since $a^2 = 2b^2 + 1$.

10.4. Numerical data

The matrix A of Proposition 10.2 is in our case equal to

$$A = \begin{pmatrix} 3 & 4 \\ 2 & 3 \end{pmatrix} \qquad \text{hence} \qquad A^2 = \begin{pmatrix} 17 & 24 \\ 12 & 17 \end{pmatrix}.$$

The greater eigenvalue of A is $3 + 2\sqrt{2}$; that of A^2 is $(3 + 2\sqrt{2})^2 = 17 + 12\sqrt{2}$. Thus, the growth exponent for the parameter a is $17 + 12\sqrt{2} \approx 33.97$. The parameter m is proportional to a^2, hence its growth exponent is

$$(3 + 2\sqrt{2})^4 = (17 + 12\sqrt{2})^2 \approx 1153.999133\ldots \tag{10.6}$$

First eight values of a divisible by 3 are

$$3,\ 99,\ 3363,\ 114\,243,\ 3\,880\,899,\ 131\,836\,323,\ 4\,478\,554\,083,\ 152\,139\,002\,499.$$

First four values of m are

$$\begin{aligned} a &= 3 && \Rightarrow && m = 2, \\ a &= 99 && \Rightarrow && m = 1634, \\ a &= 3363 && \Rightarrow && m = 1\,884\,962, \\ a &= 114\,243 && \Rightarrow && m = 2\,175\,243\,842. \end{aligned}$$

The orbit for $m = 2$ is not quadratic, hence the smallest degree m for which the quadratic orbit splits into two orbits over $\mathbb{Q}$ is $m = 1634$. For $a = 152\,139\,002\,499$ we have $m \approx 3.86 \cdot 10^{21}$.

Other examples of Diophantine invariants (but only for ordinary trees) may be found in [**LaZv-04**], Sections 2.2.4.2 and 2.2.4.3. See also Example 2.2.27, Proposition 2.2.28 and Example 2.2.31 there.

The example considered in Section 2.2.4.3 of [**LaZv-04**] is maybe the most interesting, or at least the most advanced. The combinatorial orbit considered in this example is the set of three ordinary trees of diameter 4. The center is of degree 7, and among the seven vertices which go around the center, two are of degree m and five are of degree k, $m \neq k$. Thus, the passport is $(m^2k^5, 7^1 1^{2m+5k-7}, (2m+5k)^1)$. (Check that there are indeed three trees and that the moduli field is totally real.) We write down the cubic polynomial defining the field of moduli and ask if it is possible for this polynomial to have a rational root (in such a case the combinatorial orbit would split into two Galois orbits, one defined over $\mathbb{Q}$, and another one defined over a real quadratic field). It turns out that this question is reduced to a search of rational points on an elliptic curve, one of the most classical (and advanced) topics in algebraic number theory and algebraic geometry. The rank of the curve in question is equal to 1. Therefore, it contains infinitely many rational points, so while computing them we had to stop somewhere. We have stopped after finding eleven cases of splitting. The 11th solution contains trees with approximately $3.45 \cdot 10^{134}$ edges.

EXERCISE 10.6 (Pell's equation once again). Consider the following passport: $(m^2, 5^1 1^{2m-5})$, $m \geq 5$.

(1) Draw the trees having this passport. Make sure that there are two of them, and that the corresponding field is real.
(2) Compute the Belyĭ function. The corresponding field is $\mathbb{Q}(\sqrt{\Delta})$ where $\Delta = 3(m-2)(2m-3)$.
(3) This time both $m-2$ and $2m-3$ can be divisible by 3, so we must consider two cases:

$$\text{(A)} \qquad 2m-3 = a^2, \quad m-2 = 3b^2,$$

$$\text{(B)} \qquad m-2 = a^2, \quad 2m-3 = 3b^2.$$

(4) Show that the system of equations (A) is reduced to the Pell equation

$$a^2 - 6b^2 = 1.$$

Find the fundamental solution and the general formula giving all solutions. Find several numerical solutions (a, b) and the corresponding values of m.

(5) Show that the system of equations (B) is reduced to the Pell-like equation

$$c^2 - 6b^2 = -2$$

where $c = 2a$. Find the smallest solution and, using the results for equation (A), find the general formula giving all solutions. Show that the value of the variable c for all the solutions is even. Find several numerical solutions (c, b) and the corresponding values of m.

(6) Find the growth exponents for the values of m obtained from the solutions of (A) and (B).

CHAPTER 11

Enumeration

This chapter is based on the results of the papers [**Zvo-13**] by Zvonkin and [**Koc-13**] by Kochetkov.

11.1. Enumeration according to the weight and the number of edges

In this section we formulate the main results; the proofs will follow in the subsequent sections.

DEFINITION 11.1 (Rooted tree). A tree with a distinguished edge is called *rooted tree*, and the distinguished edge itself is called its *root*. We consider the root edge as being oriented from black to white. (Note that this definition does *not* coincide with Definition 5.8 of Chapter 5, page 30.)

The goal of this chapter is the enumeration of rooted weighted plane trees.

THEOREM 11.2 (According to the weight). *Let a_n denote the number of rooted weighted bicolored plane trees of weight n. Then the generating function*

$$f(t) = \sum_{n\geq 0} a_n t^n$$

is equal to

$$\begin{aligned} f(t) &= \frac{1-t-\sqrt{1-6t+5t^2}}{2t} \\ &= 1+t+3t^2+10t^3+36t^4+137t^5+ \\ &\quad 543t^6+2219t^7+9285t^8+\dots \end{aligned} \tag{11.1}$$

This function can also be represented as a continued fraction

$$f(t) = \cfrac{1}{1-\cfrac{u}{1-\cfrac{u}{1-\cfrac{u}{1-\dots}}}} \quad \textit{where} \quad u = \frac{t}{1-t}. \tag{11.2}$$

The numbers a_n satisfy the following recurrence relations which resemble the recurrences for Motzkin numbers:

$$a_0 = 1, \quad a_1 = 1, \qquad a_{n+1} = a_n + \sum_{k=0}^{n} a_k a_{n-k} \quad \textit{for} \quad n \geq 1, \tag{11.3}$$

and

$$a_n = \frac{6n-3}{n+1}\cdot a_{n-1} - \frac{5n-10}{n+1}\cdot a_{n-2} \quad \textit{for} \quad n\geq 2. \tag{11.4}$$

The asymptotic formula for the numbers a_n is

$$a_n \sim \frac{1}{2}\sqrt{\frac{5}{\pi}} \cdot 5^n\, n^{-3/2}. \tag{11.5}$$

THEOREM 11.3 (According to the weight and the number of edges). *Let $b_{m,n}$ denote the number of rooted weighted bicolored plane trees of weight n with m edges. Then the generating function*

$$h(s,t) = \sum_{m,n\geq 0} b_{m,n} s^m t^n$$

is equal to

$$\begin{aligned} h(s,t) &= \frac{1-t-\sqrt{1-(2+4s)\,t+(1+4s)\,t^2}}{2st} \\ &= 1+st+(s+2s^2)\,t^2+(s+4s^2+5s^3)\,t^3+ \\ &\quad (s+6s^2+15s^3+14s^4)\,t^4+\ldots \end{aligned} \tag{11.6}$$

The following is an explicit formula for the numbers $b_{m,n}$:

$$b_{m,n} = \binom{n-1}{m-1} \cdot \mathrm{Cat}_m = \binom{n-1}{m-1} \cdot \frac{1}{m+1}\binom{2m}{m}, \tag{11.7}$$

where Cat_m is the m-th Catalan number. Finally, the function $h(s,t)$ admits the continued fraction representation (11.2) *with*

$$u = \frac{st}{1-t}.$$

THEOREM 11.4 (Mass-formula). *Denote $|\mathrm{Aut}(T)|$ the order of the automorphism group of a tree T. Let c_n denote the number of non-isomorphic non-rooted trees T of weight n, the contribution of a tree T to the total sum being equal to* $1/|\mathrm{Aut}(T)|$. *Then*

$$c_n = \sum_T \frac{1}{|\mathrm{Aut}(T)|} = \sum_{m=1}^{n} \frac{b_{m,n}}{m}, \tag{11.8}$$

where the first sum is taken over all the non-isomorphic non-rooted weighted trees T of weight n.

The sequence a_n is listed in the On-Line Encyclopedia of Integer Sequences [**OEIS**] as the entry A002212. It has many different interpretations, some of them coming from chemistry. Among the various interpretations there are "multi-trees" (Roland Bacher, 2005) which correspond to our weighted trees. Roland Bacher informed the authors that these trees appeared as a byproduct of his earlier topological studies. Some of the above-stated formulas may also be found in [**OEIS**].

EXAMPLE 11.5 (Trees of weight 4). Figure 11.1 shows all the trees of weight 4. There are ten trees in the picture but in fact there are 16 non-isomorphic (non-rooted) trees of weight 4. Indeed, when we exchange black and white, four trees remain isomorphic to themselves while six others do not, so we must add the six missing trees to those shown explicitly in the figure.

Near each tree, the number of its possible rootings is indicated, with color exchange taken into account. We see that the total number of trees is 36, which is the coefficient a_4 in front of t^4 in $f(t)$, see (11.1). Among these 36 trees, there is one tree with one edge, six trees with two edges, 15 trees with three edges, and 14 trees with four edges. These are the coefficients of the polynomial $s + 6s^2 + 15s^3 + 14s^4$ which stands in front of t^4 in $h(s,t)$, see (11.6).

The number c_4, according to (11.8), is equal to

$$1 + \frac{6}{2} + \frac{15}{3} + \frac{14}{4} = 12\frac{1}{2}.$$

And, indeed, among the 16 non-isomorphic non-rooted trees there are ten asymmetric trees, four trees with the symmetry of order 2, and two trees with the symmetry of order 4, which gives

$$10 + 4 \cdot \frac{1}{2} + 2 \cdot \frac{1}{4} = 12\frac{1}{2}.$$

We leave other details to the reader.

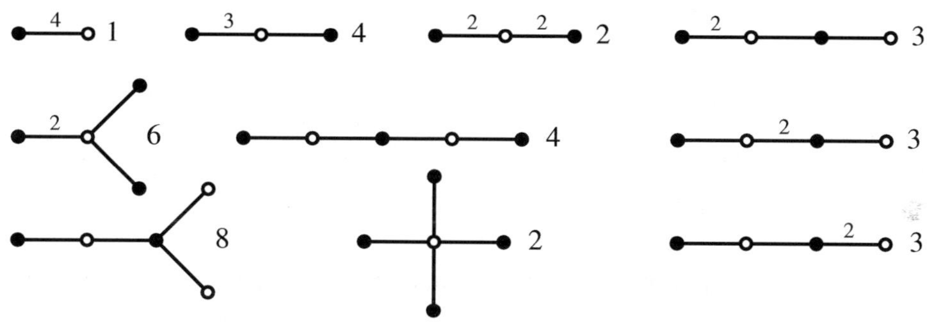

FIGURE 11.1. Near each tree, the number of its possible rootings is indicated, with an eventual color exchange taken into account. The total number of rooted trees is 36.

Proofs of the three above theorems are given in Section 11.3. Certain preliminary constructions necessary for the proof are carried out in Section 11.2.

11.2. Dyck words and weighted Dyck words

There is a standard way of encoding rooted topological (non-weighted) plane trees by *Dyck words* and *Dyck paths*. We start on the left bank of the root edge and go around the tree in the clockwise direction, writing the letter x when we follow an edge for the first time, and the letter y when we follow it the second time on its opposite side. A Dyck path corresponding to a Dyck word is a path on the plane which starts at the origin and takes a step $(1, 1)$ for every letter x and a step $(1, -1)$ for every letter y.

These objects can be easily characterized. For a word w which is a concatenation of the three words, $w = u_1u_2u_3$ (anyone of them is allowed to be empty), we call u_1 a *prefix*, u_2 a *factor*, and u_3 a *suffix* of w.

DEFINITION 11.6 (Dyck words and Dyck paths). A *Dyck word* is a word w in the alphabet $\{x, y\}$ such that $|w|_x = |w|_y$ (here $|w|_x$ and $|w|_y$ stand for the number of occurences of x and y in w), while for any prefix u of w we have $|u|_x \geq |u|_y$. A *Dyck path* is a path on the plane which starts at the origin, takes steps $(1, 1)$ and $(1, -1)$, and finishes on the horizontal axis, while always staying on the upper half-plane.

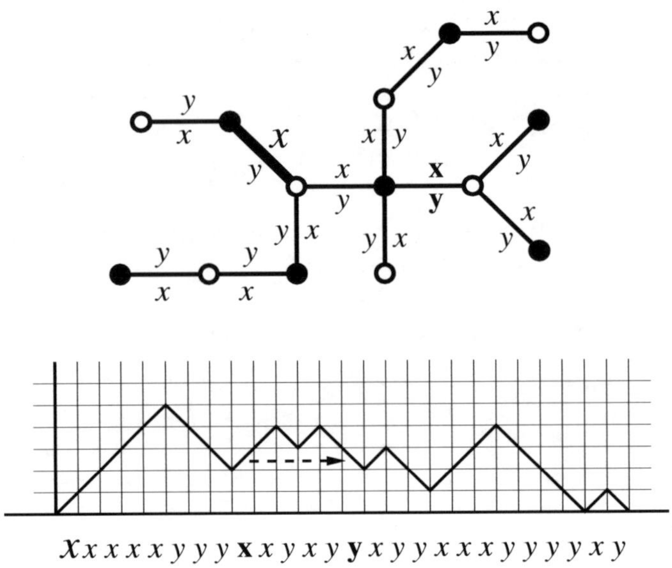

FIGURE 11.2. A rooted bicolored plane tree and its encoding by the corresponding Dyck word and Dyck path.

Figure 11.2 illustrates these notions. The root edge in the tree is shown by the thick line, while the initial letter x is shown in a bigger font.

REMARK 11.7 (Empty Dyck word). A Dyck word may be empty. Then it corresponds to the tree consisting of a single vertex. Making an exception to the general rule, we do not color this vertex in black or white. Thus, there exists a single empty word, and a single tree without edges.

The following proposition is a trivial consequence of the above construction.

PROPOSITION 11.8 (Trees and Dyck words). *There is a bijection between rooted bicolored plane trees and Dyck words.*

There remains very little to do in order to describe weighted trees. There is a natural notion of coupling the letters of a Dyck word: a *couple* is the pair of letters (x, y) standing on the opposite sides of the same edge. It is easy to recognize a couple in a word or in a path. Let x, y be a pair of letters in a Dyck word w, where x comes before y. Consider the factor xuy of w which starts with x and terminates with y. Then the pair (x, y) forms a couple if and only if the factor u between x and y is a Dyck word. In a Dyck path, we take an ascending step corresponding to a letter x and go horizontally until we meet a descending step opposite to it: this step corresponds to the letter y which forms a couple with x. In Figure 11.2, an example of a couple, both on the tree and in the word, is indicated in a boldface font, and the dashed arrow shows how to find the "opposite" step. The couple (x, y) in which x is the very first letter of the Dyck word, corresponds to the root edge.

Now, returning to the weighted trees, we do the following: for every edge of the tree, we take the corresponding couple (x, y) and replace it with (x_i, y_i) where i is the weight of the edge.

DEFINITION 11.9 (Weighted Dyck words). A *weighted Dyck word* is a word in the infinite alphabet $\{x_i, y_i\}_{i\geq 1}$ which is a Dyck word in which every couple of letters (x, y) is replaced by a certain couple (x_i, y_i). We say that a couple (x_i, y_i) has the weight i, and the weight of a word is the sum of the weights of all its couples.

PROPOSITION 11.10 (Weighted words and trees). *There is a bijection between rooted weighted bicolored plane trees and weighted Dyck words.*

11.3. Proof of the three enumerative theorems

Every non-empty Dyck word w has a unique decomposition of the form $w = xuyv$ where u and v are themselves Dyck words (maybe empty). Here, obviously, x is the first letter of w, and y is the letter coupled with it. The step in the Dyck path corresponding to the letter y is the descending step of the first return of the path to the horizontal axis.

In the same way, every non-empty *weighted* Dyck word w has a unique decomposition of the form $x_i u y_i v$ for some $i \geq 1$, where u and v are weighted Dyck words.

Let $\mathcal{D}$ be the formal sum of all the weighted Dyck words, that is, the formal power series

$$\begin{aligned} \mathcal{D} \;=\;& \varepsilon + x_1y_1 + x_2y_2 + x_1x_1y_1y_1 + x_1y_1x_1y_1 + x_3y_3 + x_1x_2y_2y_1 + \\ & x_1y_1x_2y_2 + x_2x_1y_1y_2 + x_2y_2x_1y_1 + x_1x_1x_1y_1y_1y_1 + \ldots \end{aligned} \tag{11.9}$$

in non-commuting variables x_i, y_i, $i = 1, 2, \ldots$, where ε stands for the empty word. (In order to write down this series we must choose a total order on the set of words in the alphabet $\{x_i, y_i\}_{i\geq 1}$. A particular choice of the order is irrelevant. In (11.9), the words are ordered, first, by their weight, then, for a given weight, by the number of steps, and then, for a given weight and number of steps, in the alphabetic order.) Then, the above decomposition of the words of $\mathcal{D}$ in the form $x_i u y_i v$ implies the following equation for $\mathcal{D}$:

$$\mathcal{D} \;=\; \varepsilon + x_1\mathcal{D}y_1\mathcal{D} + x_2\mathcal{D}y_2\mathcal{D} + \ldots \;=\; \varepsilon + \sum_{i=1}^{\infty} x_i\mathcal{D}y_i\mathcal{D}\,. \tag{11.10}$$

Now, let us do the following:

- replace ε and each occurrence of a letter y_i in $\mathcal{D}$ by a factor 1;
- replace each occurrence of a letter x_i in $\mathcal{D}$ by $s\,t^i$;
- make the variables s and t commute.

Then, every word w in $\mathcal{D}$ is transformed into a word $s^m t^n$ where m is the number of occurrences in w of the letters x_i, $i \geq 1$, taken together (or, equivalently, the number of edges of the weighted tree T_w corresponding to w), and n is the weight of w (or, equivalently, the total weight of T_w). Therefore, combining similar terms we get the generating function $h(s,t) = \sum_{m,n\geq 0} b_{m,n} s^m t^n$. At the same time, equation (11.10) is transformed into the following quadratic equation for $h(s,t)$:

$$h \;=\; 1 + s\left(\sum_{i=1}^{\infty} t^i\right)h^2 \;=\; 1 + \frac{st}{1-t}\cdot h^2. \tag{11.11}$$

Solving this equation, and choosing the sign in front of the square root in such a way as to avoid a pole at zero, we obtain formula (11.6). Then, substituting $s = 1$ in (11.6) we get (11.1).

In order to obtain the asymptotic expression (11.5) for the numbers a_n it suffices to apply to $f(t)$ the ready-made formulas of asymptotic analysis of the coefficients of generating functions, see, for example, Chapter VI of the book **[FlSe-09]**. The only thing to note is that

$$1 - 6t + 5t^2 = (1-t)(1-5t).$$

In order to prove (11.7), we proceed as follows. There are Cat_m topological rooted trees with m edges. Starting at the root edge, we go around a tree in the clockwise direction and attribute a non-zero weight to every newly encountered edge. There are $\binom{n-1}{m-1}$ ways to do that. Indeed, put n dots in a row, and distribute $m-1$ separators among $n-1$ places between the dots. This procedure splits the number n into m non-zero parts.

The continued fraction (11.2) implies the equality

$$f = \frac{1}{1 - u \cdot f}.$$

This equality is a quadratic equation with respect to f. Solving it, and taking the appropriate branch, we get

$$f = \frac{1 - \sqrt{1-4u}}{2u},$$

which is, by the way, the generating function for the Catalan numbers if we consider u as a formal parameter. Substituting now

$$u = \frac{t}{1-t}$$

we get the expression (11.1) for f, and taking

$$u = \frac{st}{1-t}$$

we return to the equation (11.11).

In order to prove the recurrence (11.3), consider separately the trees of weight $n+1$ having the root edge of weight 1, and the trees of weight $n+1$ having the root edge of weight $i \geq 2$. The weighted Dyck words corresponding to the trees of the first kind are of the form $x_1 u y_1 v$, where u and v are themselves weighted Dyck words. The sum of the weights of u and v is n; denoting the weight of u by k, so that the weight of v becomes $n-k$, and summing over the $k = 0, 1, \ldots, n$, we get the term $\sum_{k=0}^{n} a_k a_{n-k}$ of (11.3). Now, all the trees of weight $n+1$ having the root edge of weight $i \geq 2$ are obtained from the trees of weight n having the root edge of weight $i-1$, by adding one unit to the weight of the root. This gives the term a_n in the right-hand part of (11.3).

Finally, (11.8) follows from the fact that there are m choices of a root edge in a tree T with m edges, but if this tree has non-trivial symmetries then some of these choices produce isomorphic rooted trees. The number of non-isomorphic rootings is $m/|\mathrm{Aut}(T)|$. Thus, dividing by m, we get the factor $1/|\mathrm{Aut}(T)|$.

In order to prove the recurrence (11.4), let us introduce the differential operator $D = t \cdot d/dt$. This operator multiplies the coefficient in front of t^n by n. Then the equality

$$(n+1)a_n = (6n-3)a_{n-1} - (5n-10)a_{n-2} \quad \mathit{for} \quad n \geq 2$$

is equivalent to the following one:

$$(D+1)f = t \cdot (6D+3)f - 5t^2 \cdot Df + (A + Bt),$$

where f is the generating function (11.1). The term $A + Bt$ appears since a_0 and a_1 do not satisfy the recurrence. The latter equality can be verified either by hand or with the help of Maple, and, indeed, it is true, with $A + Bt = 1 - t$.

All the three theorems of Section 11.1 are proved. □

11.4. Enumeration according to a passport

Here we follow the paper [**Koc-13**] by Kochetkov. At certain places our presentation becomes rather informal.

Let $\alpha, \beta \vdash n$ be two partitions of n, and let p, q be the numbers of parts in α and β respectively.

DEFINITION 11.11 (Non-separable passport). We call a pair of partitions (α, β) of n a *valuable passport* if $p + q \leq n + 1$. A passport is *separable* if we can subdivide both α and β into two subsets, $\alpha = \alpha_1 \sqcup \alpha_2$, $\beta = \beta_1 \sqcup \beta_2$, such that both (α_1, β_1) and (α_2, β_2) are valuable passports. If such a subdivision is impossible then the passport (α, β) is called *non-separable.*

EXAMPLE 11.12 (Separable and non-separable passports). Let us take $n = 12$ and consider the following passport: $(5^1 4^1 3^1, 2^5 1^2)$. It is separable. Indeed, it can be subdivided into three valuable passports: $(5^1, 2^2 1^1) \sqcup (4^1, 2^2) \sqcup (3^1, 2^1 1^1)$. The passport $(5^1 4^1 3^1, 6^2)$ is non-separable, as well as the passport $(9^5, 5^9)$.

The problem with separable passports is that they correspond to trees but also to forests, see Figure 11.3, while our goal is to enumerate only trees. Therefore, some kind of an inclusion-exclusion procedure may turn out to be inevitable. However, for ordinary trees there exists a direct formula which does not use an inclusion-exclusion. We will discuss this case a little later. Thus, not all hopes are as yet lost to get a direct formula for weighted trees.

For non-separable passports this sort of difficulties does not arise. Therefore, we will deal mainly with the non-separable case.

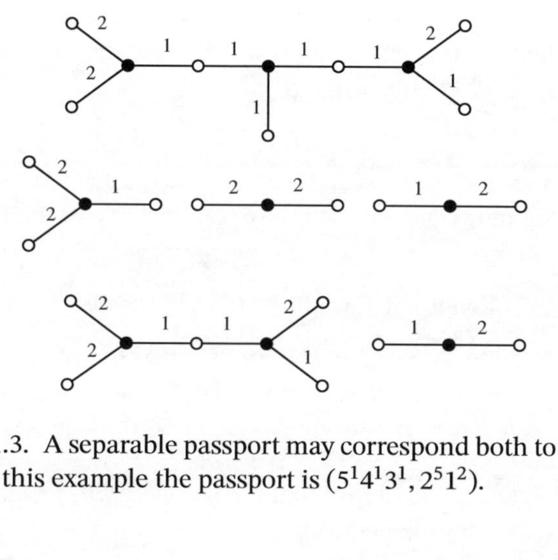

FIGURE 11.3. A separable passport may correspond both to trees and forests. In this example the passport is $(5^1 4^1 3^1, 2^5 1^2)$.

DEFINITION 11.13 (Totally labelled tree). Let a tree be given. Fix a color and a degree, and suppose that there are k vertices of this type. Then we label them by numbers $1, 2, \dots, k$. When we apply this procedure independently to all colors and degrees of the vertices of the tree we get a *totally labelled tree.*

THEOREM 11.14 (Enumeration of totally labelled trees). *Let (α, β) be a non-separable passport, and the numbers of parts in α and β be p and q respectively. Then the number of totally labelled trees with this passport is $(p + q - 2)!$.*

NOTATION 11.15 (Factorial of a partition). Let $\lambda \vdash n$ be a partition of n. Consider the "power notation" $\lambda = 1^{d_1} 2^{d_2} \dots n^{d_n}$. Then we denote

$$\lambda! = d_1! \, d_2! \dots d_n!$$

COROLLARY 11.16 (Enumeration of non-labelled trees). *Let (α, β) be a non-separable passport, and the numbers of parts in α and β be p and q respectively. Then*

$$\sum \frac{1}{|\mathrm{Aut}(T)|} = \frac{(p+q-2)!}{\alpha! \cdot \beta!}. \tag{11.12}$$

where the sum is taken over the trees with the passport (α, β).

EXAMPLE 11.17 (Counting non-labelled trees with non-separable passports). For the passport $(9^5, 5^9)$ from Example 9.21 we have $p = 5, q = 9, p+q-2 = 12$. Therefore, the number of trees having this passport is

$$\frac{12!}{5! \cdot 9!} = \frac{10 \cdot 11 \cdot 12}{5!} = 11.$$

For the passport $(5^1 4^1 3^1, 6^2)$ from Example 11.12 the number of trees is $3!/2! = 3$.

For the passport $(6^1, 2^2 1^2)$ our formula gives

$$\frac{3!}{2! \cdot 2!} = \frac{3}{2} = 1 + \frac{1}{2},$$

and, indeed, there are two trees corresponding to this passport: one of them is asymmetric while the other one is symmetric, with the symmetry of order 2.

For the passport $(4^3 3^1, 5^3)$ the formula gives

$$\frac{5!}{3! \cdot 3!} = \frac{120}{36} = 3 + \frac{1}{3}$$

and, indeed, there are four trees with this passport and one of then is symmetric with the symmetry of order 3.

EXERCISE 11.18 (Drawing). Draw all the trees of the above example.

REMARK 11.19 (Ordinary trees). One might get an impression that the passports corresponding to ordinary trees are non-separable. Indeed, for a tree with n edges the number of vertices is $p + q = n + 1$ while a forest with n edges will have more than $n + 1$ vertices.

But this is to forget the possibility to use the edges with a weight greater than 1. Figure 11.4 shows what may take place.

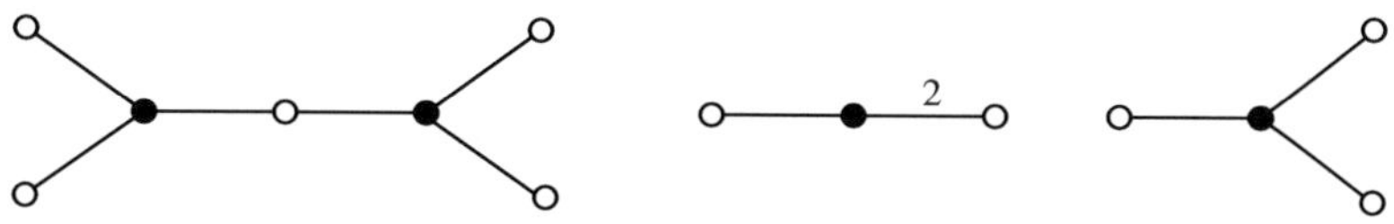

FIGURE 11.4. An ordinary tree with the passport $(3^2, 2^1 1^4)$, and a forest with the same passport. The forest contains five edges instead of six, and there is an edge with the weight greater than 1.

In spite of that, there is a formula due to Tutte [**Tut-64**] which counts the ordinary trees without using inclusion-exclusion. Namely,

$$\sum \frac{1}{|\mathrm{Aut}(T)|} = \frac{(p-1)!\cdot(q-1)!}{\alpha!\cdot\beta!} \tag{11.13}$$

where the sum is taken over the ordinary trees with the passport (α, β). Comparing this formula to (11.12) we see that (11.13) is obtained from (11.12) by dividing it by the binomial coefficient

$$\binom{p+q-2}{p-1} = \binom{p+q-2}{q-1}.$$

We don't have an explicit explanation of this fact.

PROOF OF THEOREM 11.14. The main idea is as follows. The black, respectively white, vertices of a tree are roots of the polynomial P, respectively Q, of the DZ-pair corresponding to the tree. Both P and Q are of degree n, while their difference R is of degree $(n+1)-(p+q)$ where p and q are numbers of black, respectively white, vertices. Thus, the number of coefficients in P and Q equal to each other is

$$n - [(n+1)-(p+q)] = p+q-1.$$

In fact, we can always make leading coefficients equal to 1, while the remaining $p+q-2$ equalities are algebraic conditions on the roots of P and Q. Being more specific, we notice that the coefficients of P and Q are elementary symmetric functions of their roots. The degrees of these functions are $1, 2, \ldots, p+q-2$. Therefore, equating them we get a system of algebraic equations of degree $(p+q-2)!$. A little trouble is with the number of the unknowns which is $p+q$, that is, greater than the number $p+q-2$ of the equation. Hence, we have to add to the system two arbitrary *linear* equations. For example, we may fix the values of two variables (i. e., positions of two vertices).

What remains to do is to establish the following facts:

- The system is non-degenerate, so that all its solutions are distinct.
- There are no solutions disappearing at infinity.
- We should also figure out why this scheme does not work for separable passports. What exactly goes wrong with the above system of algebraic equations for separable passports?

In order to simplify notation let us denote $p+q-2 = m$. It is also convenient to pass from the elementary symmetric functions to the power functions. Then our system will take the form

$$\begin{aligned}
\alpha_1 x_1 + \alpha_2 x_2 + \ldots + \alpha_p x_p &= \beta_1 y_1 + \beta_2 y_2 + \ldots + \beta_q y_q, \\
\alpha_1 x_1^2 + \alpha_2 x_2^2 + \ldots + \alpha_p x_p^2 &= \beta_1 y_1^2 + \beta_2 y_2^2 + \ldots + \beta_q y_q^2, \\
\ldots\ldots\ldots & \quad \ldots\ldots\ldots \\
\alpha_1 x_1^m + \alpha_2 x_2^m + \ldots + \alpha_p x_p^m &= \beta_1 y_1^m + \beta_2 y_2^m + \ldots + \beta_q y_q^m.
\end{aligned}$$

Once again, we simplify the notation. Fixing the values of $x_p = 0$ and $y_q = 1$; denoting $\alpha_1, \ldots, \alpha_{p-1}, -\beta_1, \ldots, -\beta_{q-1}$ by $\lambda_1, \ldots, \lambda_m$; denoting $x_1, \ldots, x_{p-1}, y_1, \ldots, y_{q-1}$ by $z_1, \ldots, z_m$, we get the following system:

$$(11.14) \qquad \sum_{k=1}^{m} \lambda_k z_k = \beta_q, \qquad \sum_{k=1}^{m} \lambda_k z_k^2 = \beta_q, \qquad \ldots, \qquad \sum_{k=1}^{m} \lambda_k z_k^m = \beta_q.$$

The matrix of derivatives of these equations with respect to z_k, $k = 1, \ldots, m$ looks as follows:

$$D_z = \begin{pmatrix} \lambda_1 & \lambda_2 & \ldots & \lambda_m \\ 2\lambda_1 z_1 & 2\lambda_2 z_2 & \ldots & 2\lambda_m z_m \\ \ldots & \ldots & \ldots & \ldots \\ m\lambda_1 z_1^{m-1} & m\lambda_2 z_2^{m-1} & \ldots & m\lambda_m z_m^{m-1} \end{pmatrix}.$$

The determinant of this matrix is obviously equal to

$$\det D_z = m! \cdot \Delta(z_1, \ldots, z_m) \cdot \prod_{k=1}^{m} \lambda_k$$

where $\Delta(z_1, \ldots, z_m)$ is the Vandermonde determinant.

In Chapter 3 we have seen that two vertices of the same color cannot coincide since in such a case the lower bound of the degree of $P - Q$ would not be attained. But why is it impossible to obtain a parasitic solution in which a black vertex would coincide with a white one? The answer is, in such a case the polynomials P and Q of the DZ-pair would not be coprime: $P = P_1 \cdot T$, $Q = Q_1 \cdot T$ with a non-constant polynomial T. Not coprime, so what? We are speaking of a parasitic solution, aren't we? The answer is, in such a case the passport of the pair (P, Q) would be separable: its sub-passport corresponding to the pair (P_1, Q_1) can be separated from the other parts of the passport of (P, Q).

We conclude that the Vandermonde determinant $\Delta(z_1, \ldots, z_m)$ cannot be equal to zero, so the above matrix D_z is non-degenerate.

Now, the matrix of the derivatives of equations (11.14) by λ_k, $k = 1, \ldots, m$ is equal to

$$D_\lambda = \begin{pmatrix} z_1 & z_2 & \ldots & z_m \\ z_1^2 & z_2^2 & \ldots & z_m^2 \\ \ldots & \ldots & \ldots & \ldots \\ z_1^m & z_2^m & \ldots & z_m^m \end{pmatrix},$$

and its determinant is equal to

$$\det D_\lambda = \Delta(z_1, \ldots, z_m) \cdot \prod_{k=1}^{m} z_k.$$

Recalling the formulas of the derivation of implicit functions we conclude that the Jacobian of the function

$$h : (\lambda_1, \dots, \lambda_m) \mapsto (z_1, \dots, z_m)$$

is equal to

$$-\frac{\det D_\lambda}{\det D_z} = -\frac{\prod_{k=1}^{m} z_k}{m! \prod_{k=1}^{m} \lambda_k}.$$

We have previously fixed $x_p = 0$; therefore, no other z_k is equal to zero. We conclude that the function h is non-degenerate, and we cannot obtain the same tree twice.

To have a solution at infinity means to consider our variables as projective ones, and the solution itself should satisfy the system similar to (11.14) but with zeros instead of β_q at the right-hand side:

$$\sum_{k=1}^{m} \lambda_k z_k = 0, \qquad \sum_{k=1}^{m} \lambda_k z_k^2 = 0, \qquad \dots, \qquad \sum_{k=1}^{m} \lambda_k z_k^m = 0. \tag{11.15}$$

We consider (11.15) as a linear system with the unknowns $\lambda_1, \dots, \lambda_m$. We know that it has a non-zero solution; hence, the determinant $D_\lambda = 0$, so that there exist some variables $z_k = 0$ or $z_i - z_j = 0$. Then, in the case $z_k = 0$ we eliminate all the terms $\lambda_k z_k^l$, $l = 1, \dots, m$ from the system, and in the case $z_i - z_j = 0$ we replace z_j with z_i everywhere and adjust the coefficients replacing λ_j with $\lambda_i + \lambda_j$. Since all the variables z_k cannot be zero we will have to stop at some moment. This means that our passport is separable.

Theorem 11.14, and hence its Corollary 11.16, are proved. □

The inclusion-exclusion formula for separable passports is not very useful, so we will formulate it without proof. But first we will illustrate it by an example. The idea of the proof is rather attractive. After writing down the system (11.14) we remark that it makes sense not necessarily for integer coefficients but also for non-integer ones. By a small perturbation of a passport we may transform any passport into a non-separable one.

EXAMPLE 11.20 (A small perturbation of a separable passport). The passport $\pi = (5^1 3^1, 5^1 3^1)$ is separable. Let us change it in the following way:

$$\pi_\varepsilon = (5^1 3^1, (5-\varepsilon)^1 (3+\varepsilon)^1).$$

The passport π_ε is non-separable, and, indeed, there are two trees corresponding to it: see Figure 11.5.

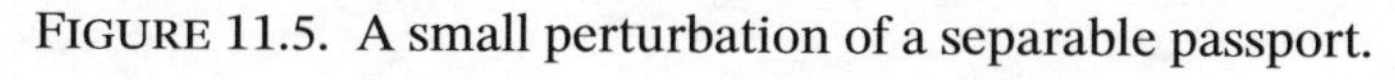

FIGURE 11.5. A small perturbation of a separable passport.

We see that when ε becomes equal to zero the tree on the right splits into two. The proof of the theorem below proceeds by counting the numbers of edges of an "infinitely small" weight and by eliminating the corresponding trees.

EXERCISE 11.21 (ε-edges). Take the passport of Figure 11.3 (page 169); perturb it in order to obtain a non-separable passport and draw the trees with one and with two ε-edges (and also without ε-edges).

THEOREM 11.22 (Inclusion-exclusion). *Let $v(\pi)$ denote the number of vertices of a tree with the passport π: namely, if $\pi = (\alpha, \beta)$ and $\alpha = (\alpha_1, \ldots, \alpha_p)$, $\beta = (\beta_1, \ldots, \beta_q)$ then $v(\pi) = p + q$. Suppose that a passport π can be represented as a union of r passports: $\pi = \pi_1 \cup \ldots \cup \pi_r$. Then the contribution of this union to the number of totally labelled* (connected!) *trees is*

$$(-1)^{r-1}(v(\pi) - 1)^{r-2} \prod_{i=1}^{r} (v(\pi_i) - 1)!$$

The reader may consider this statement as an exercise or find a proof in Kochetkov's paper [**Koc-13**].

CHAPTER 12

What remains to be done

In this chapter we collect several open questions, problems and directions of possible future research. Their level of difficulty is variable: some of them are clearly feasible, for the other ones it is not that obvious. Some questions are rigorously stated, some remain rather vague. It may turn out that some of our problems are devoid of meaning: the future will show... The order of presentation of the problems makes no difference.

IN THE STEPS OF BEUKERS AND STEWART [**BeSt-10**]. It would be interesting to draw the trees corresponding to all the series of polynomials and all the sporadic polynomials of their paper. It would be even more interesting to find *combinatorially* something new which did not appear in their paper. A good subject for an undergraduate diploma work.

DIOPHANTINE INVARIANTS. This is certainly a vast and exciting subject. Only, it is very difficult to formulate a concrete question or to state a concrete problem. One must engage in a sort of a blind search, hoping to discover something interesting and beautiful. Concerning, in particular, Pell's equations, we may propose two questions: (a) simulate *all* Pell and/or Pell-like equations by dessins d'enfants; and (b) find combinatorial orbits which *never* split into more than one Galois orbit. A culminating result of this line of research would be a discovery of an algorithmically undecidable problems in the theory of dessins d'enfants (if such problems exist).

Since a positive outcome in studying Diophantine invariants is not guaranteed, we can recommend this direction of research only to a senior researcher whose career is already well established.

NEGATIVE DEGREES. When working with parametric series of dessins, we may formally attribute negative values to degrees of certain vertices. Geometrically, it may mean, for example, that a parameter which used to be the degree of a black vertex becomes the degree of a white one (or vice versa). At the same time, all formal expressions for Belyĭ functions remain valid, and the algebraic form of solutions remains the same. This allows one to extend a given series to a larger set of examples or unite several series into one. In fact, we have seen in this book several such examples though we did not attract a special attention to them. See, for example, the generalized Jacobi polynomials with parameters $a, b < -1$, Section 6.5: sometimes we almost have a whim to say that the black degrees are positive while the white ones are negative. This line of research may lead to very interesting observations.

ENUMERATION ACCORDING TO A PASSPORT. Find an explicit formula giving the number of weighted trees with a given separable passport without resorting to inclusion-exclusion. Find a bijective proof of the formula for non-separable passports.

INVERSE ENUMERATION PROBLEM. Classify combinatorial orbits of sizes 2 and 3. In [**Adr-07a**], it is proved that the combinatorial orbits of size $k \geq 2$ of ordinary trees are classified as follows: there are several infinite series of trees of diameter ≤ 6, and a finite number of sporadic trees of diameter $\leq 12k + 1$. Find an analogue of this result for the weighted trees.

FACTORIZATION OF DISCRIMINANTS. It was first found experimentally by Zvonkin, and then proved by Birch [**Bir-07**], that for ordinary trees a discriminant of a polynomial, whose splitting field is the moduli field of a Galois orbit, factorizes into linear combinations of vertex degrees. A similar pattern seems to be valid for weighted trees. Formulate the corresponding statement and prove it. To give but one example: for ordinary trees the biggest factor of the discriminant is the number of edges. Incidentally, this number is equal to the degree of the outer face. For weighted trees this equality is no longer valid, and in numerous examples it is the degree of the outer face which takes part in the factorization: see, for example, the factor $3m - 2$ in the expression (10.3) or the factor $2m - 3$ in Exercise 10.6. The knowledge of the factorization may give rise to some concrete and explicit Diophantine problems. See also the remarks made on page 58: certain factors of the discriminant can be guessed by purely geometric observations. By the way, negative degrees may become very useful in this context.

MODULI FIELDS FOR SPECIAL TREES. Compute the moduli fields for the remaining cases of Chapter 8.

MODULI FIELDS AND CHARACTERS. Specialists in the inverse Galois theory (see [**MaMa-18**], [**Ser-92**]) certainly can explain close relations between moduli fields and characters of the monodromy groups of dessins that we have observed in Chapter 8. As to us, what we would like to point out is that the values of the characters appear in the final result as if out of the blue: we know of no algorithms of computing of Belyĭ functions which would use characters in any explicit way. It may turn out that computation of characters covers a large part of a preliminary work needed in order to compute Belyĭ functions, so that by using characters the computation of Belyĭ functions would be significantly simplified.

When the monodromy group is S_n, all entries of the character tables belong to $\mathbb{Q}$ while only a tiny minority of dessins with this monodromy group are defined over $\mathbb{Q}$. Still, it may turn out that the knowledge of the characters of S_n may largely simplify the equations we need to solve in order to find Belyĭ functions and the corresponding number fields.

This remark concerns not only weighted trees but arbitrary dessins.

COMPOSITION AND DECOMPOSITION OF WEIGHTED TREES. To find a combinatorial and algorithmic counterpart of the operation of composition of Belyĭ functions, or, more generally, of ramified coverings, is far from being an easy task. However, it becomes rather simple for ordinary plane trees: cf. [**AdZv-98**]. We suppose that it should also be reasonably simple for weighted trees.

There are, essentially, only two families of polynomials admitting non-unique decomposition. These are power monomials x^n and Chebyshev polynomials $T_n(x)$. Both facts are easily observable using ordinary trees. For rational functions this simple pattern is no longer valid. Can weighted trees shed a new light on this subject?

MEGAMAPS. The fact that being a megamap is a Galois invariant is part of a "common knowledge" but it seems that its proof was never written in a clear and explicit way. As we have already mentioned in Chapter 9, we don't have an algorithm which, for a given map, would say if it is a megamap for some Fried family. Even when we do know that a given map is a megamap, we don't know how to find the corresponding Fried family (or families). One more question: can we say, for a given passport, if there are megamaps among the maps having this passport?

We are sure that a profound study of this subject will bring about more and more interesting questions about megamaps.

CONCERNING THE *abc* CONJECTURE: DON'T CHERISH VAIN HOPES. This last remark is of a negative nature. It is well known that Belyĭ functions help to construct interesting examples of *abc*-triples. The specificity of the Belyĭ functions treated in this book is that the degrees of finite faces are all equal to one. However, if we have a face of degree k with its center at x_0, then it is not $(x-x_0)^k$ but just $x-x_0$ which is one of the factors of the radical. Thus, faces of degree one are no better than faces of any other degree. At the same time, the number of faces is determined by the number of vertices and edges. Therefore, DZ-pairs, with their faces of degree one, do not give any advantage as compared to other Belyĭ functions.

Bibliography

[AbSt-72] M. Abramowitz, I. Stegun, eds., *Handbook of mathematical functions: with formulas, graphs, and mathematical tables*, Dover, 1972.

[Adr-97a] N. M. Adrianov, Arithmetic Theory of Graphs on Surfaces (Russian), Ph. D. Thesis, Moscow State University, 1997, 116 pp.

[Adr-97b] N. M. Adrianov, *Classification of primitive edge rotation groups of plane trees* (Russian, with English and Russian summaries), Fundam. Prikl. Mat. **3** (1997), no. 4, 1069–1083. MR1794502

[Adr-07a] N. M. Adrianov, *On planar trees with a prescribed number of valency set realizations* (Russian, with English and Russian summaries), Fundam. Prikl. Mat. **13** (2007), no. 6, 9–17, DOI 10.1007/s10958-009-9369-3; English transl., J. Math. Sci. (N.Y.) **158** (2009), no. 1, 5–10. MR2476026

[Adr-07b] N. M. Adrianov, *On generalized Chebyshev polynomials corresponding to planar trees of diameter 4* (Russian, with English and Russian summaries), Fundam. Prikl. Mat. **13** (2007), no. 6, 19–33, DOI 10.1007/s10958-009-9371-9; English transl., J. Math. Sci. (N.Y.) **158** (2009), no. 1, 11–21. MR2476027

[Adr-16] N. M. Adrianov, *Primitive monodromy groups of rational functions with one multiple pole*, Zap. Nauchn. Sem. S.-Peterburg. Otdel. Mat. Inst. Steklov. (POMI) **446** (2016), no. Kombinatorika i Teoriya Grafov. V, 12–30, DOI 10.1007/s10958-017-3549-3; English transl., J. Math. Sci. (N.Y.) **226** (2017), no. 5, 548–560. MR3520420

[AKS-97] N. M. Adrianov, Yu. Yu. Kochetkov, and A. D. Suvorov, *Plane trees with exceptional primitive edge rotation groups* (Russian, with English and Russian summaries), Fundam. Prikl. Mat. **3** (1997), no. 4, 1085–1092. MR1794503

[AdSh-97] N. M. Adrianov and G. B. Shabat, *Unicellular cartography and Galois orbits of plane trees*, Geometric Galois actions, 2, London Math. Soc. Lecture Note Ser., vol. 243, Cambridge Univ. Press, Cambridge, 1997, pp. 13–24, DOI 10.1017/CBO9780511666124.003. MR1653007

[AdZv-98] N. M. Adrianov and A. K. Zvonkin, *Composition of plane trees*, Acta Appl. Math. **52** (1998), no. 1-3, 239–245, DOI 10.1023/A:1005971411271. Algebra and combinatorics: interactions and applications (Königstein, 1994). MR1649700

[AdZv-13] N. M. Adrianov and A. K. Zvonkin, *Weighted trees with primitive edge rotation groups* (Russian, with English and Russian summaries), Fundam. Prikl. Mat. **18** (2013), no. 6, 5–50, DOI 10.1007/s10958-015-2494-2; English transl., J. Math. Sci. (N.Y.) **209** (2015), no. 2, 160–191. MR3431854

[Bab-89] L. Babai, *The probability of generating the symmetric group*, J. Combin. Theory Ser. A **52** (1989), no. 1, 148–153, DOI 10.1016/0097-3165(89)90068-X. MR1008166

[Bar-03] E. J. Barbeau, *Pell's equation*, Problem Books in Mathematics, Springer-Verlag, New York, 2003. MR1949691

[Bel-79] G. V. Belyĭ, *Galois extensions of a maximal cyclotomic field* (Russian), Izv. Akad. Nauk SSSR Ser. Mat. **43** (1979), no. 2, 267–276, 479. English transl., Math. USSR Izvestiya **14** (1980), 247–256. MR534593

[BeSt-10] F. Beukers and C. L. Stewart, *Neighboring powers*, J. Number Theory **130** (2010), no. 3, 660–679, DOI 10.1016/j.jnt.2009.09.006. MR2584847

[BBM-14] Y. F. Bilu, Y. Bugeaud, and M. Mignotte, *The problem of Catalan*, Springer, Cham, 2014. MR3288807

[Bir-07] B. J. Birch, *Shabat trees of diameter 4: attachment to a letter of Zvonkin* (Russian, with English and Russian summaries), Fundam. Prikl. Mat. **13** (2007), no. 6, 131–135, DOI 10.1007/s10958-009-9379-1; English transl., J. Math. Sci. (N.Y.) **158** (2009), no. 1, 94–96. MR2476031

[BCHS-65] B. J. Birch, S. Chowla, M. Hall Jr., and A. Schinzel, *On the difference* $x^3 - y^2$, Norske Vid. Selsk. Forh. (Trondheim) **38** (1965), 65–69. MR186620

[Boc-82] G. Boccara, *Cycles comme produit de deux permutations de classes données* (French), Discrete Math. **38** (1982), no. 2-3, 129–142, DOI 10.1016/0012-365X(82)90282-5. MR676530

[BoSt-04] P. L. Bowers and K. Stephenson, *Uniformizing dessins and Belyĭ maps via circle packing*, Mem. Amer. Math. Soc. **170** (2004), no. 805, xii+97, DOI 10.1090/memo/0805. MR2053391

[Bug-10] V. O. Bugaenko, Pell's Equation (Russian), Moscow, MCCME, 2010.

[But-93] G. Butler, *The transitive groups of degree fourteen and fifteen*, J. Symbolic Comput. **16** (1993), no. 5, 413–422, DOI 10.1006/jsco.1993.1056. MR1271082

[BMK-83] G. Butler and J. McKay, *The transitive groups of degree up to eleven*, Comm. Algebra **11** (1983), no. 8, 863–911, DOI 10.1080/00927878308822884. MR695893

[CNT-82] P. J. Cameron, P. M. Neumann, and D. N. Teague, *On the degrees of primitive permutation groups*, Math. Z. **180** (1982), no. 2, 141–149, DOI 10.1007/BF01318900. MR661693

[CaCo-99] P. Cassou-Noguès and J.-M. Couveignes, *Factorisations explicites de* $g(y) - h(z)$ (French), Acta Arith. **87** (1999), no. 4, 291–317, DOI 10.4064/aa-87-4-291-317. MR1671641

[Chu-70] G. V. Čudnovskiĭ, *Diophantine predicates* (Russian), Uspehi Mat. Nauk **25** (1970), no. 4 (154), 185–186. MR0276172

[Cor-75] R. Cori, *Un code pour les graphes planaires et ses applications*, Société Mathématique de France, Paris, 1975. With an English abstract; Astérisque, No. 27. MR0404045

[CPZ-12] P. Corvaja, C. Petronio, and U. Zannier, *On certain permutation groups and sums of two squares* (English, with German summary), Elem. Math. **67** (2012), no. 4, 169–181, DOI 10.4171/EM/207. MR2990143

[Cou-94] J.-M. Couveignes, *Calcul et rationalité de fonctions de Belyĭ en genre* 0 (French, with English and French summaries), Ann. Inst. Fourier (Grenoble) **44** (1994), no. 1, 1–38. MR1262878

[Dan-82] L. V. Danilov, *The Diophantine equation* $x^3 - y^2 = k$ *and a conjecture of M. Hall* (Russian), Mat. Zametki **32** (1982), no. 3, 273–275, 425. MR677595

[Dan-89] L. V. Danilov, *The Diophantine equations* $x^m - Ay^n = k$ (Russian), Mat. Zametki **46** (1989), no. 6, 38–45, 126, DOI 10.1007/BF01158625; English transl., Math. Notes **46** (1989), no. 5-6, 914–919 (1990). MR1051048

[Dav-65] H. Davenport, *On* $f^3(t) - g^2(t)$, Norske Vid. Selsk. Forh. (Trondheim) **38** (1965), 86–87. MR186621

[DDH-89] S. Diaz, R. Donagi, and D. Harbater, *Every curve is a Hurwitz space*, Duke Math. J. **59** (1989), no. 3, 737–746, DOI 10.1215/S0012-7094-89-05933-4. MR1046746

[Duj-10] A. Dujella, *On Hall's conjecture*, Acta Arith. **147** (2011), no. 4, 397–402, DOI 10.4064/aa147-4-5. MR2776096

[EKS-84] A. L. Edmonds, R. S. Kulkarni, and R. E. Stong, *Realizability of branched coverings of surfaces*, Trans. Amer. Math. Soc. **282** (1984), no. 2, 773–790, DOI 10.2307/1999265. MR732119

[Elk-00] N. D. Elkies, *Rational points near curves and small nonzero* $|x^3 - y^2|$ *via lattice reduction*, Algorithmic number theory (Leiden, 2000), Lecture Notes in Comput. Sci., vol. 1838, Springer, Berlin, 2000, pp. 33–63, DOI 10.1007/10722028_2. MR1850598

[Elk-13] N. D. Elkies, *The complex polynomials* $P(x)$ *with* $\mathrm{Gal}(P(x) - t) \cong M_{23}$, ANTS X—Proceedings of the Tenth Algorithmic Number Theory Symposium, Open Book Ser., vol. 1, Math. Sci. Publ., Berkeley, CA, 2013, pp. 359–367, DOI 10.2140/obs.2013.1.359. MR3207422

[ElkHall] N. D. Elkies, *List of integers* x, y, *with* $x < 10^{18}$, $0 < |x^3 - y^2| < x^{1/2}$, `http://www.math.harvard.edu/~elkies/hall.html`.

[FlSe-09] P. Flajolet and R. Sedgewick, *Analytic combinatorics*, Cambridge University Press, Cambridge, 2009. MR2483235

[Fri-77] M. Fried, *Fields of definition of function fields and Hurwitz families—groups as Galois groups*, Comm. Algebra **5** (1977), no. 1, 17–82, DOI 10.1080/00927877708822158. MR453746

[FGM-14] D. Frohardt, R. Guralnick, and K. Magaard, *Primitive monodromy groups of genus at most two*, J. Algebra **417** (2014), 234–274, DOI 10.1016/j.jalgebra.2014.06.020. MR3244646

[FrMa-01] D. Frohardt and K. Magaard, *Composition factors of monodromy groups*, Ann. of Math. (2) **154** (2001), no. 2, 327–345, DOI 10.2307/3062099. MR1865973

[Gal-DB] A database of Galois groups, `http://www.lmfdb.org/GaloisGroup/`.

[GiGo-12] E. Girondo and G. González-Diez, *Introduction to compact Riemann surfaces and dessins d'enfants*, London Mathematical Society Student Texts, vol. 79, Cambridge University Press, Cambridge, 2012. MR2895884

[GoJa-92] I. P. Goulden and D. M. Jackson, *The combinatorial relationship between trees, cacti and certain connection coefficients for the symmetric group*, European J. Combin. **13** (1992), no. 5, 357–365, DOI 10.1016/S0195-6698(05)80015-0. MR1181077

[Gro-84] A. Grothendieck, *Esquisse d'un programme* (French, with French summary), Geometric Galois actions, 1, London Math. Soc. Lecture Note Ser., vol. 242, Cambridge Univ. Press, Cambridge, 1997, pp. 5–48. With an English translation on pp. 243–283. MR1483107

[GuSh-07] R. M. Guralnick and J. Shareshian, *Symmetric and alternating groups as monodromy groups of Riemann surfaces. I. Generic covers and covers with many branch points*, Mem. Amer. Math. Soc. **189** (2007), no. 886, vi+128, DOI 10.1090/memo/0886. With an appendix by Guralnick and R. Stafford. MR2343794

[GuTh-90] R. M. Guralnick and J. G. Thompson, *Finite groups of genus zero*, J. Algebra **131** (1990), no. 1, 303–341, DOI 10.1016/0021-8693(90)90178-Q. MR1055011

[Hal-71] M. Hall Jr., *The Diophantine equation* $x^3 - y^2 = k$, Computers in number theory (Proc. Sci. Res. Council Atlas Sympos. No. 2, Oxford, 1969), Academic Press, London, 1971, pp. 173–198. MR0323705

[Hul-05] A. Hulpke, *Constructing transitive permutation groups*, J. Symbolic Comput. **39** (2005), no. 1, 1–30, DOI 10.1016/j.jsc.2004.08.002. MR2168238

[Hur-1891] A. Hurwitz, *Ueber Riemann'sche Flächen mit gegebenen Verzweigungspunkten* (German), Math. Ann. **39** (1891), no. 1, 1–60, DOI 10.1007/BF01199469. MR1510692

[Hus-62] D. H. Husemoller, *Ramified coverings of Riemann surfaces*, Duke Math. J. **29** (1962), 167–174. MR136726

[JaWi-09] M. J. Jacobson Jr. and H. C. Williams, *Solving the Pell equation*, CMS Books in Mathematics/Ouvrages de Mathématiques de la SMC, Springer, New York, 2009. MR2466979

[Jon-14] G. A. Jones, *Primitive permutation groups containing a cycle*, Bull. Aust. Math. Soc. **89** (2014), no. 1, 159–165, DOI 10.1017/S000497271300049X. MR3163014

[JoWo-16] G. A. Jones and J. Wolfart, *Dessins d'enfants on Riemann surfaces*, Springer Monographs in Mathematics, Springer, Cham, 2016. MR3467692

[JoZv-02] G. A. Jones and A. K. Zvonkin, *Orbits of braid groups on cacti* (English, with English and Russian summaries), Mosc. Math. J. **2** (2002), no. 1, 127–160, 200, DOI 10.17323/1609-4514-2002-2-1-127-160. MR1900588

[JoMa-91] J. P. Jones and Y. V. Matijasevič, *Proof of recursive unsolvability of Hilbert's tenth problem*, Amer. Math. Monthly **98** (1991), no. 8, 689–709, DOI 10.2307/2324421. MR1130680

[Jor-1870] C. Jordan, *Traité des substitutions et des équations algébriques* (French), Les Grands Classiques Gauthier-Villars. [Gauthier-Villars Great Classics], Éditions Jacques Gabay, Sceaux, 1989. Reprint of the 1870 original. MR1188877

[Jor-1871] C. Jordan, *Mémoire sur les groupes primitifs* (French), Bull. Soc. Math. France **1** (1872/73), 175–221. MR1503659

[Koc-13] Yu. Yu. Kochetkov, *Enumeration of a class of plane weighted trees* (Russian, with English and Russian summaries), Fundam. Prikl. Mat. **18** (2013), no. 6, 171–184, DOI 10.1007/s10958-015-2503-5; English transl., J. Math. Sci. (N.Y.) **209** (2015), no. 2, 282–291. MR3431863

[LJB-66] M. Lal, M. F. Jones, and W. J. Blundon, *Numerical solutions of the Diophantine equation* $y^3 - x^2 = k$, Math. Comp. **20** (1966), 322–325, DOI 10.2307/2003511. MR191871

[LaZv-04] S. K. Lando, A. K. Zvonkin, *Graphs on surfaces and their applications*, with an appendix by D. B. Zagier. Encyclopaedia of Mathematical Sciences, vol. 141, Low dimensional topology, II, Springer-Verlag, Berlin, 2004. See also ERRATA AND COMMENTS to this book: `https://www.labri.fr/perso/zvonkin/Books/errata.pdf`

[Lan-90] S. Lang, *Old and new conjectured Diophantine inequalities*, Bull. Amer. Math. Soc. (N.S.) **23** (1990), no. 1, 37–75, DOI 10.1090/S0273-0979-1990-15899-9. MR1005184

[MSW-11] K. Magaard, S. Shpectorov, and G. Wang, *Generating sets of affine groups of low genus*, Computational algebraic and analytic geometry, Contemp. Math., vol. 572, Amer. Math. Soc., Providence, RI, 2012, pp. 173–192, DOI 10.1090/conm/572/11366. MR2953829

[Mag-97] N. Magot, Cartes planaires et fonctions de Belyĭ : aspects algorithmiques et expérimentaux, Ph. D. thesis, Université Bordeaux I, 1997, 144 pp.

[MaMa-18] G. Malle and B. H. Matzat, *Inverse Galois theory*, Springer Monographs in Mathematics, Springer, Berlin, 2018. Second edition [MR1711577]. MR3822366

[Mat-96] Yu. V. Matiyasevich, *Computation of generalized Chebyshev polynomials on a computer* (Russian, with Russian summary), Vestnik Moskov. Univ. Ser. I Mat. Mekh. **6** (1996), 59–61, 112; English transl., Moscow Univ. Math. Bull. **51** (1996), no. 6, 39–40 (1997). MR1481512

[Mat-98] Yu. V. Matiyasevich, *Generalized Chebyshev polynomials*, Personal Journal, 1998, `http://logic.pdmi.ras.ru/~yumat/personaljournal/chebyshev/chebysh.htm`.

[Mat-15] Yu. V. Matiyasevich, Private communication, March 2015 – April 2016.

[Mat-16] Yu. Matiyasevich, *Calculation of Belyĭ functions for trees with weighted edges*, Zap. Nauchn. Sem. S.-Peterburg. Otdel. Mat. Inst. Steklov. (POMI) **446** (2016), no. Kombinatorika i Teoriya Grafov. V, 122–138, DOI 10.1007/s10958-017-3554-6; English transl., J. Math. Sci. (N.Y.) **226** (2017), no. 5, 623–634. MR3520425

[Mih-04] P. Mihăilescu, *Primary cyclotomic units and a proof of Catalan's conjecture*, J. Reine Angew. Math. **572** (2004), 167–195, DOI 10.1515/crll.2004.048. MR2076124

[Mon-14] H. Monien, *How to calculate rational coverings for subgroups of* $\mathrm{PSL}_2(\mathbb{Z})$ *efficiently*, Slides of the talk given at the conference "Embedded Graphs", St. Petersburg, October 2014.

[Mul-95] P. Müller, *Primitive monodromy groups of polynomials*, Recent developments in the inverse Galois problem (Seattle, WA, 1993), Contemp. Math., vol. 186, Amer. Math. Soc., Providence, RI, 1995, pp. 385–401, DOI 10.1090/conm/186/02193. MR1352284

[Mul-15] P. Müller, *A combined Gröbner basis and power series approach in inverse Galois theory*, Slides of the talk given at the conference "Constructive Methods in Number Theory", Bethe Center for Theoretical Physics, Bonn, March 2015.

[OEIS] The On-Line Encyclopedia of Integer Sequences, `http://oeis.org/`.

[Oga-13] D. A. Oganesyan, *On rational functions with two critical points of maximum multiplicity* (Russian, with English and Russian summaries), Fundam. Prikl. Mat. **18** (2013), no. 6, 185–208, DOI 10.1007/s10958-015-2504-4; English transl., J. Math. Sci. (N.Y.) **209** (2015), no. 2, 292–308. MR3431864

[Oga-16] D. Oganesyan, *Abel pairs and modular curves*, Zap. Nauchn. Sem. S.-Peterburg. Otdel. Mat. Inst. Steklov. (POMI) **446** (2016), no. Kombinatorika i Teoriya Grafov. V, 165–181, DOI 10.1007/s10958-017-3556-4; English transl., J. Math. Sci. (N.Y.) **226** (2017), no. 5, 655–666. MR3520427

[Ore-03] S. Yu. Orevkov, *Riemann existence theorem and construction of real algebraic curves* (English, with English and French summaries), Ann. Fac. Sci. Toulouse Math. (6) **12** (2003), no. 4, 517–531. MR2060598

[Pak-09] F. Pakovich, *Solution of the Hurwitz problem for Laurent polynomials*, J. Knot Theory Ramifications **18** (2009), no. 2, 271–302, DOI 10.1142/S0218216509006896. MR2507927

[PaZv-14] F. Pakovich and A. K. Zvonkin, *Minimum degree of the difference of two polynomials over* $\mathbb{Q}$*, and weighted plane trees*, Selecta Math. (N.S.) **20** (2014), no. 4, 1003–1065, DOI 10.1007/s00029-014-0151-0. MR3273629

[PaZv-18] F. Pakovich and A. K. Zvonkin, *Davenport-Zannier polynomials over* $\mathbb{Q}$, Int. J. Number Theory **14** (2018), no. 4, 925–974, DOI 10.1142/S1793042118500562. MR3801075

[PaPe-09] M. A. Pascali and C. Petronio, *Surface branched covers and geometric 2-orbifolds*, Trans. Amer. Math. Soc. **361** (2009), no. 11, 5885–5920, DOI 10.1090/S0002-9947-09-04779-5. MR2529918

[PellWi] *Pell's equation*, Wikipedia, `https://en.wikipedia.org/wiki/Pell%27s_equation`.

[PellSol] `http://www.jakebakermaths.org.uk/maths/jshtmlpellsolverbigintegerv10.html` : Pell's solver.

[Rin-74] G. Ringel, *Map color theorem*, Springer-Verlag, New York-Heidelberg, 1974. Die Grundlehren der mathematischen Wissenschaften, Band 209. MR0349461

[Rob-14] D. P. Roberts, *Chebyshev covers and exceptional number fields*, Preprint, 2014, 22 pp. `http://cda.mrs.umn.edu/~roberts/dpr/Research_files/CCaENF.pdf`

[Rob-15] D. P. Roberts, *Some Belyĭ covers unexpectedly (from a purely 3-point viewpoint) defined over* $\mathbb{Q}$, Slides of the talk given at the conference "Constructive Methods in Number Theory", Bethe Center for Theoretical Physics, Bonn, March 2015.

[Rob-16] D. P. Roberts, *Hurwitz-Belyi maps* (English, with English and French summaries), Publications mathématiques de Besançon. Algèbre et théorie des nombres. 2018, Publ. Math. Besançon Algèbre Théorie Nr., vol. 2018, Presses Univ. Franche-Comté, Besançon, 2018, pp. 25–67. MR3838687

[Sch-94] L. Schneps (ed.), *The Grothendieck theory of dessins d'enfants*, London Mathematical Society Lecture Note Series, vol. 200, Cambridge University Press, Cambridge, 1994. Papers from the Conference on Dessins d'Enfant held in Luminy, April 19–24, 1993. MR1305390

[Ser-92] J.-P. Serre, *Topics in Galois theory*, Research Notes in Mathematics, vol. 1, Jones and Bartlett Publishers, Boston, MA, 1992. Lecture notes prepared by Henri Damon [Henri Darmon]; With a foreword by Darmon and the author. MR1162313

[Sha-19] G. B. Shabat, *Belyĭ pairs in the critical filtrations of Hurwitz spaces*, `https://arxiv.org/pdf/1902.02034.pdf`.

[Shi-05] T. Shioda, *Elliptic surfaces and Davenport-Stothers triples*, Comment. Math. Univ. St. Pauli **54** (2005), no. 1, 49–68. MR2153955

[SiVo-14] J. Sijsling and J. Voight, *On computing Belyi maps* (English, with English and French summaries), Numéro consacré au trimestre "Méthodes arithmétiques et applications", automne 2013, Publ. Math. Besançon Algèbre Théorie Nr., vol. 2014/1, Presses Univ. Franche-Comté, Besançon, 2014, pp. 73–131. MR3362631

[SiVo-16] J. Sijsling and J. Voight, *On explicit descent of marked curves and maps*, Res. Number Theory **2** (2016), Art. 27, 35, DOI 10.1007/s40993-016-0057-3. MR3582054

[Sto-81] W. W. Stothers, *Polynomial identities and Hauptmoduln*, Quart. J. Math. Oxford Ser. (2) **32** (1981), no. 127, 349–370, DOI 10.1093/qmath/32.3.349. MR625647

[Sze-39] G. Szegő, *Orthogonal Polynomials*, American Mathematical Society Colloquium Publications, v. 23, American Mathematical Society, New York, 1939. Reprinted in 1992. MR0000077

[Tho-65] R. Thom, *L'équivalence d'une fonction différentiable et d'un polynome* (French), Topology **3** (1965), no. suppl, suppl. 2, 297–307, DOI 10.1016/0040-9383(65)90079-0. MR187249

[Tut-64] W. T. Tutte, *The number of planted plane trees with a given partition*, Amer. Math. Monthly **71** (1964), 272–277, DOI 10.2307/2312183. MR186585

[Vat-19] A. Vatuzov, Private communication, September 2019.

[Voi-16] J. Voight, Private communication, February 2016.

[Wal-75] T. R. S. Walsh, *Hypermaps versus bipartite maps*, J. Combinatorial Theory Ser. B **18** (1975), 155–163, DOI 10.1016/0095-8956(75)90042-8. MR360328

[Wan-11] G. Wang, Genus Zero Systems for Primitive Groups of Affine Type, Ph. D. Thesis, University of Birmingham, 2011, 98 pp.

[Zan-95] U. Zannier, *On Davenport's bound for the degree of* $f^3 - g^2$ *and Riemann's existence theorem*, Acta Arith. **71** (1995), no. 2, 107–137, DOI 10.4064/aa-71-2-107-137. MR1339121

[Zan-01] U. Zannier, *Good reduction of certain covers* $\mathbf{P}^1 \to \mathbf{P}^1$, Israel J. Math. **124** (2001), 93–114, DOI 10.1007/BF02772609. MR1856506

[Zvo-01] A. K. Zvonkin, *Megamaps: construction and examples*, Discrete models: combinatorics, computation, and geometry (Paris, 2001), Discrete Math. Theor. Comput. Sci. Proc., AA, Maison Inform. Math. Discrèt. (MIMD), Paris, 2001, pp. 329–339. MR1888783

[Zvo-13] A. K. Zvonkin, *Enumeration of weighted plane trees* (Russian, with English and Russian summaries), Fundam. Prikl. Mat. **18** (2013), no. 6, 135–144, DOI 10.1007/s10958-015-2500-8; English transl., J. Math. Sci. (N.Y.) **209** (2015), no. 2, 258–264. MR3431860

Index